Contemporary Diagnosis and Management of

LIPID DISORDERS®

Antoni… Phil

Dear…

Weill Med… University,

New York, NY

Third Edition

Best Regards
Antonio M. Gotto, Jr.

Published by
Handbooks in Health Care Co.,
Newtown, Pennsylvania, USA

This book is presented as a service to the medical community and is intended for informational and educational use only. It is not a substitute for individualized treatment based on a careful assessment and diagnosis of each patient. The opinions are those of the author and are not attributable to the publisher or the provider of the educational grant. As stated below, the author is a consultant to, or board member of, several pharmaceutical and medical device companies that manufacture and/or sell products related to the medical conditions described in this book.

The parties involved in the writing and publication of this book are not responsible for the application of its content in clinical practice. Information on drug indications, adverse effects, drug-drug interactions, and dosing is based on recommendations current at the time of publication. However, this book is not intended to replace the complete prescribing information for each drug. Reference to the complete prescribing information is essential before any drug is used in patients. Each physician must also exercise clinical judgment in evaluating the benefits of pharmacologic treatment against the risk(s) of toxicity. In addition, this book is not intended to replace a careful reading of the complete guidelines issued by the American Heart Association, the National Cholesterol Education Program Adult Treatment Panel III, the American Diabetes Association, and other authoritative bodies.

Author's disclosure information: Consultant to: AstraZeneca, Bristol-Myers Squibb, Kowa, Merck, Merck/Schering-Plough, Novartis, Pfizer (past 12 months), Reliant; Member of Board of Directors: Medtronic

International Standard Book Number: 1-931981-18-3

Library of Congress Catalog Card Number: 2003109732

 For information, write Handbooks in Health Care, 3 Terry Drive, Suite 201, Newtown, Pennsylvania 18940, (215) 860-9600.
Web site: www.HHCbooks.com

Table of Contents

Introduction 5

Chapter 1
**Rationale for
the Management of Dyslipidemia** 7

Chapter 2
**Fundamentals of
Blood Lipid Metabolism
and Concepts in Atherogenesis** 43

Chapter 3
**Risk Assessment
and Risk Factor Reduction** 75

Chapter 4
**Therapeutic Options: Dietary
and Other Nondrug Interventions** 116

Chapter 5
**Therapeutic Options:
Pharmacologic Interventions** 164

Chapter 6
Special Populations 215

Appendix A: Case Reports 254

**Appendix B:
Strategies to Promote Adherence** 274

Appendix C: Compliance Pledge 276

Index 278

Acknowledgments

I am extremely grateful to three of my colleagues for their invaluable and gracious assistance in preparing the Third Edition of *Contemporary Diagnosis and Management of Lipid Disorders®*. Richard C. Pasternak, MD, Vice President, Clinical Research, Cardiovascular/Atherosclerosis, Merck & Co., Inc., carefully reviewed and updated the treatment algorithm (Chapter 3), which he prepared for the prior edition, to ensure that it reflects the revised National Cholesterol Education Program recommendations and their application in clinical practice. Carl J. Vaughan, MD, FRCPI, Visiting Assistant Professor of Medicine, Weill Medical College of Cornell University, and Consultant Cardiologist, Mercy University Hospital, Cork, Ireland, thoroughly reviewed the case reports, revising them based on the new guidelines and ensuring that each case represents the realities of patient care. Dr. Vaughan prepared the case reports for the prior edition and has provided new material for this edition. Stuart D. Saal, MD, Medical Director of the Transplantation Program and Co-Director of the Renal Consultation Service and the Maurice R. Greenberg Comprehensive Lipid Control Center, the Rogosin Institute, New York, Weill Cornell Medical Center, made an essential contribution to the chapter on pharmacologic interventions.

In addition, I would like to express my appreciation to all of my colleagues who contributed their time and experience to the prior editions of this book and whose expertise continues to enrich the current edition.

Introduction

Coronary heart disease (CHD) is the leading cause of death in the United States. The course of this insidious disease is lifelong. It begins in childhood or adolescence, when unwise eating habits and–increasingly–a sedentary lifestyle lead to dyslipidemia and the development of atherosclerotic lesions that progress over time. Other life-habit risk factors (for example, cigarette smoking and obesity) can contribute to the process. Genetic factors, including familial hypercholesterolemia, can also play an important role. For years, most people are unaware that they have atherosclerotic disease until they either experience a major coronary event or, in more fortunate cases, are tested and diagnosed before a heart attack (or stroke) occurs.

The absence of symptoms over a period of decades is a principal reason why coronary heart disease is so difficult to prevent and treat. For example, both physicians and their patients tend to overlook the use of simple diagnostic tools, such as a fasting blood lipid profile, as part of routine medical care.

This book is intended to help overcome barriers to diagnosis, treatment, and prevention. Written for physicians, it is a practical guide to risk reduction, with emphasis on the National Cholesterol Education Program's revised evidence-based guidelines for detecting, evaluating, and treating high blood cholesterol in adults. In addition to discussing the use of the guidelines in clinical practice, the book reviews the major trials on which these recommendations are based. Both pharmacologic and nonpharmacologic strategies are examined in the following pages; a chapter on fundamental concepts in blood lipid metabolism and atherogenesis is included as well.

The 20th century was a time of major advances in understanding lipid metabolism, establishing a link between high cholesterol and atherosclerotic cardiovascular disease, demonstrating that cholesterol reduction can lower cardiovascular risk, educating the public about lifestyle changes needed to prevent or slow the progression of atherosclerosis, and developing drug treatments that can reduce the incidence of major coronary events by approximately one third.

As we embark on a new century, several priorities urgently call for our attention. We must intensify public lifestyle-education efforts, particularly in the face of a growing obesity epidemic. In research, there are two areas of importance: more well-designed trials evaluating the effects of lipid modification on clinical outcomes in a variety of populations, and laboratory investigations, both basic and translational, that can lead to the development of novel drug therapies. The periodic refinement of prevention and treatment guidelines is also essential, particularly in the light of new clinical trial evidence that has been scrupulously evaluated. Finally, we must seek more effective ways of promoting physician and patient adherence to current guidelines.

Physicians are on the front line of the struggle to reduce the incidence of atherosclerotic cardiovascular disease. I hope this book will prove useful as they face the challenges of implementing risk-reduction strategies in daily clinical practice.

Antonio M. Gotto, Jr, MD, DPhil
New York, NY
November 2004

Rationale for the Management of Dyslipidemia

In the last 50 years, energetic scrutiny of the relationship between the dyslipidemias and the risk for coronary heart disease (CHD) has yielded valuable insights into the origins and management of this widespread condition. Coronary atherosclerosis is characterized by the development of sclerotic lesions in the intima and inner media of the coronary arteries. Such lesions, which consist of cholesterol and other lipoid deposits, may cause potentially severe blockage (stenosis) of the vessel or they may rupture, resulting in the formation of a blood clot that obstructs the vessel. The evolution of the atherosclerotic plaque is lifelong; the fatty streak, the earliest atherosclerotic lesion, has been found in children as young as 10 years.

The clinical consequences of atherosclerosis are far-reaching: CHD is the most common cause of death in the United States in both men and women, accounting for approximately 670,000 deaths per year.[1] Despite this statistic, reducing CHD risk with aggressive treatment of lipid disorders may be readily achieved. Lipid management is a critical component of CHD prevention. The goal of this book is not only to help the physician identify strategies for diagnosing and treating lipid disorders, but also to examine the rationale behind such approaches.

The blood lipids included in the clinical evaluation of patient risk for CHD are low-density lipoprotein cholesterol

(LDL-C), total cholesterol (TC), high-density lipoprotein cholesterol (HDL-C), and triglyceride (TG).[2] In patients with TG levels ≥200 mg/dL, non-HDL-C should also be determined. Non-HDL-C can be defined as TC minus HDL-C or as LDL-C plus very-low-density lipoprotein cholesterol (VLDL-C). Non-HDL-C is intended to capture not only LDL-C, but also TG-rich remnant lipoproteins. Atherogenic remnant lipoproteins, which are included in VLDL-C, are the most readily available measure of TG in clinical practice.[2] Chapter 2 provides a brief overview of blood lipid metabolism and the possible mechanisms of atherogenesis.

The Adult Treatment Panel III (ATP III) guidelines of the US National Cholesterol Education Program (NCEP) focus on reducing CHD risk by managing hypercholesterolemia, primarily elevated concentrations of LDL-C.[2,3] For patients with TG levels ≥200 mg/dL, a secondary therapeutic target is the lowering of non-HDL-C. The ATP III guidelines differentiate patients according to four CHD risk categories: CHD or CHD risk equivalents, ≥2 major risk factors with a 10-year CHD risk of 10% to 20%, ≥2 risk factors with a 10-year risk of <10%, and 0-1 risk factor with a 10-year CHD risk of <10%.[3] Ten-year risk is estimated by using a modified risk scoring system developed from the Framingham Heart Study (Chapter 3). Coronary heart disease risk equivalents include other clinical forms of atherosclerotic disease, diabetes, and the presence of ≥2 major risk factors with a 10-year CHD risk >20%. The benefit of cholesterol-lowering therapy in high-risk individuals without diagnosed coronary disease has recently been substantiated by the results of the Heart Protection Study (HPS).[4]

A key goal of the NCEP and other public health organizations is to refine risk identification and treatment at the individual level. The adoption of widespread screening programs and educational programs for patients and physicians is crucial to reducing CHD morbidity and mortality.

Treatment of hypercholesterolemia is the strategy outlined in this book, although management of other lipid disorders, such as low HDL-C and elevated fasting serum TG, is also addressed.

Decisions about the management of lipid disorders must be based on a careful evaluation of the patient's global risk for CHD and on the identification of underlying disorders that may preclude a diagnosis of primary dyslipidemia. Chapter 3 covers risk assessment, including Frederickson phenotyping and the major risk factors described by ATP III. For each CHD risk category, the ATP III guidelines specify LDL-C cut points for initiating therapeutic lifestyle changes (TLC) or drug therapy and the lipid goal to be achieved. Chapter 4 examines TLC, which are essential for patients receiving and not receiving lipid-modifying drug therapy. In Chapter 5, drug interventions and considerations for their use are addressed, while Chapter 6 reviews issues regarding at-risk patients in groups that, until recently, were underrepresented in clinical trials. Appendix A presents a series of case histories.

Rationale for the Management of Dyslipidemia

The predictive relationship between plasma cholesterol levels and the risk for CHD has been well established by numerous observational and clinical trials. The observational Framingham Heart Study found a positive association between CHD incidence and TC levels in women aged 35 to 94 years and men aged 35 to 64 years.[5] The Framingham study was also instrumental in identifying other major CHD risk factors, such as hypertension, cigarette smoking, and elevated blood glucose levels. Data from the Multiple Risk Factor Intervention Trial (MRFIT) demonstrate a strong, graded, positive correlation between serum cholesterol levels and CHD mortality rate (Figure 1-1).[6] The Johns Hopkins Precursor Study showed that cholesterol concentrations measured in men aged 20 to 29 years predicted later development of CHD.[7] The progressive, multifactorial, and lifelong etiology of atherosclerotic disease is borne out by data from the Bogalusa Heart Study, in which the extent of early atherosclerotic lesions was associated with the number of risk

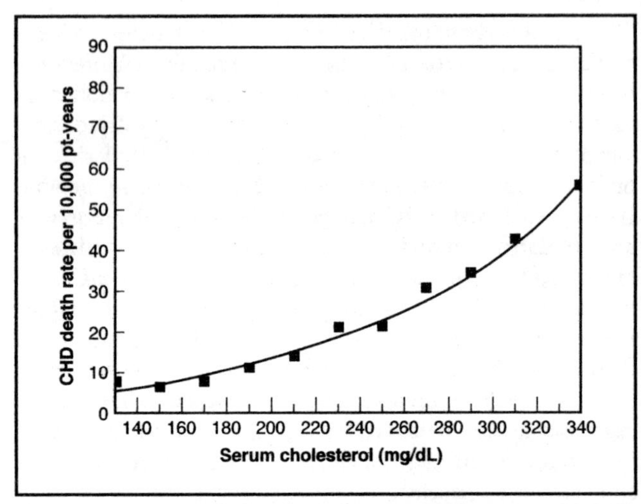

Figure 1-1: Age-adjusted CHD mortality rate per 10,000 patient-years according to serum cholesterol concentration in 316,099 patients in the Multiple Risk Factor Intervention Trial. Serum cholesterol was measured once at screening and patients were given no instructions on fasting. (Adapted and reprinted with permission from Neaton JD, Wentworth D: Serum cholesterol, blood pressure, cigarette smoking, and death from coronary heart disease. Overall findings and differences by age for 316,099 white men. Multiple Risk Factor Intervention Trial Research Group. *Arch Intern Med* 1992;152:56-64).

factors present in young adults,[8] and from the Cardiovascular Risk in Young Finns study[9] and the Pathobiological Determinants of Atherosclerosis in Youth (PDAY) study.[10,11]

As the evidence linking lipids to atherosclerosis risk strengthened, the 'lipid hypothesis' emerged, postulating that lowering serum cholesterol, whether through dietary and other lifestyle modifications alone, or through lifestyle changes in combination with cholesterol-lowering pharmacotherapy, would decrease the incidence of CHD events.

The Lipid Research Clinics Coronary Primary Prevention Trial (LRC-CPPT) tested the lipid hypothesis in 3,806 asymptomatic men with primary hypercholesterolemia who were randomized to receive 24 g/d of the bile-acid sequestrant (or 'resin') cholestyramine (LoCholest®, Questran®, Prevalite®; described in greater detail in Chapter 5) or placebo; the mean follow-up was 7.4 years.[12,13] All patients received a lipid-lowering diet. Entry criteria included an LDL-C concentration of 190 mg/dL or more and a TC level of at least 265 mg/dL. Resin therapy reduced TC and LDL-C by 13% and 20%, respectively, compared with dietary therapy, which resulted in reductions of 5% and 8%. Cholestyramine treatment was associated with a 19% reduction in CHD events, defined as a combination of nonfatal myocardial infarction (MI) and CHD death ($P<0.05$). The LRC-CPPT clearly established the clinical benefits of cholesterol lowering. In general, a 1% reduction in TC led to a 2% to 3% reduction in CHD risk.

In the 5-year primary-prevention Helsinki Heart Study (HHS), 4,081 asymptomatic men with non-HDL-C levels of 200 mg/dL or more were randomized to receive either 600 mg of the fibric acid derivative gemfibrozil (Lopid®; Chapter 5) twice daily or placebo.[14] Definite MI (fatal and nonfatal), sudden cardiac death, or unwitnessed deaths from CHD were reduced by 34% ($P<0.02$). In a refined analysis of these results, patients with the combination of elevated LDL-C and relatively low HDL-C levels (ie, LDL-C/HDL-C ratio >5) and elevated TG levels (>200 mg/dL) derived the most benefit from gemfibrozil therapy (71% risk reduction).[15]

Evidence of Clinical Benefit From the Statin Trials

Some early trials reported higher rates of noncardiovascular mortality in subjects receiving active treatment, which raised the controversial issue of whether lipid-lowering might increase the number of noncardiovascular deaths. In the LRC-CPPT, all-cause mortality in the cholestyramine group decreased by only a nonsignificant 7%. This reflected an excess number of violent and accidental deaths that appeared to be a chance occurrence. Analysis of long-term follow-up data on the subjects who received nicotinic acid (Niacor®) in the Coronary Drug Project (CDP) showed an 11% decrease in the overall mortality rate compared with the placebo group (*P*=0.0004) after 15 years. The study drug was discontinued almost 9 years before the end of the follow-up period.[16] The smaller Stockholm Ischaemic Heart Disease Secondary Prevention Study reported a lower rate of all-cause mortality in post-MI patients treated with a combination of nicotinic acid and clofibrate (82 deaths/276 in the control group vs 61/279 in the intervention group, $P<0.05$).[17] However, benefit seemed to be limited to patients with TG levels >133 mg/dL and was most pronounced in patients whose TG levels fell by 30% or more.

Two large-scale clinical studies in the 1990s provided evidence for the safety of lipid lowering in high-risk primary and secondary prevention. The Scandinavian Simvastatin Survival Study (4S), a randomized trial, showed that lipid lowering was associated with a significant reduction in total mortality, the only primary end point (Table 1-1).[18] Over a median 5.4-year follow-up, 4,444 men and women, aged 35 to 70, with a history of MI or angina pectoris and with TC levels between 213 and 309 mg/dL, were treated either with 20 mg/d of simvastatin (Zocor®; Chapter 5), a 3-hydroxy-3-methylglutaryl coenzyme A (HMG-CoA) reductase inhibitor (ie, 'statin'), or with placebo. Simvastatin dosage was titrated over the course of the study to reduce TC to between 116 and 201 mg/dL; thus, 37% of patients in the simvastatin group received 40 mg/d and two patients received only 10 mg/d. A 26% reduction in TC and a 36% reduction in LDL-C in the simvastatin group (compared with placebo) coincided with a 30% reduction in the risk for dying of any cause (*P*=0.0003). Subanalysis of the reasons for death showed no significant difference between simvastatin and placebo in the number of deaths from non-CHD causes (67/2,223 in the placebo group vs 71/2,221 in the simvastatin group), thereby establishing that the primary effect of simvastatin therapy on mortality was in reducing the number of CHD deaths (189 coronary deaths in the placebo group vs 111 in the simvastatin group, a 42% risk reduction). Simvastatin therapy was also associated with a 34% reduction in the risk for one or more major coronary events ($P<0.00001$) and with a 37% reduction in the risk for revascularization procedures. Coronary heart disease benefit was found across all quartiles of baseline total, LDL, and HDL cholesterol, with similar benefit in each quartile.[19]

In the West of Scotland Coronary Prevention Study (WOSCOPS), 6,595 men, 45 to 64 years of age, with no history of MI, were randomized to receive 40 mg/d of pravastatin (Pravachol®; Chapter 5) or placebo (see Table 1-1).[20] Five percent (5%) had evidence of angina pectoris. Compared with placebo, pravastatin therapy led to a 20% reduction in total plasma cholesterol and a 26% reduction in LDL-C. This study also reported significant reductions in clinical events. The risk for nonfatal MI or CHD death was reduced 31% ($P<0.001$) and suspected deaths from CHD were reduced 33% ($P=0.042$). The all-cause mortality rate was reduced by 22% ($P=0.051$). Together with the results of 4S, the WOSCOPS results alleviated many concerns about the safety of lipid lowering as an appropriate strategy for managing CHD risk in primary or secondary prevention. The Prospective Pravastatin Pooling (PPP) Project, examined below, has provided additional strong evidence concerning the safety of lipid-lowering therapy with statins.[21,22]

Benefit in Patients With 'Average' Cholesterol Concentrations

In 4S and WOSCOPS, patients were at relatively high risk for CHD either from preexisting symptomatic disease or from severe elevations in total and LDL cholesterol. Whether benefit could be obtained in individuals with mild to moderate elevations in total and LDL cholesterol, a group representing most of the population who develop CHD, was less clearly established. The Cholesterol and Recurrent Events (CARE) trial, the Long-Term Intervention with Pravastatin in Ischemic

Disease (LIPID) trial, and the Air Force/Texas Coronary Atherosclerosis Prevention Study (AFCAPS/TexCAPS) provided compelling evidence that such benefit may be substantial (Table 1-1). The following recent trials also demonstrate the benefit of lipid lowering in persons with relatively low LDL-C levels: the HPS in high-risk subjects with or without established CHD, the Anglo-Scandinavian Cardiac Outcomes Trial-Lipid Lowering Arm (ASCOT-LLA) in hypertensive patients, the Pravastatin or Atorvastatin Evaluation and Infection Therapy (PROVE-IT) trial in patients after an acute coronary syndrome (ACS), and the Collaborative Atorvastatin Diabetes Study (CARDS). These trials will be examined in subsequent sections.

The secondary-prevention CARE trial enrolled 4,159 men and women with a history of MI. Subjects were treated with either pravastatin 40 mg/d or placebo.[23] The mean LDL-C concentration in CARE was 139 mg/dL; pravastatin treatment reduced LDL-C levels by 32% from baseline and maintained mean levels of 97 to 98 mg/dL. After a mean follow-up of 5 years, pravastatin-treated patients had a 24% lower risk for a recurrent event (defined as either nonfatal MI or CHD death, $P=0.003$). Also, the benefit was greater among women than men (46% vs 20% risk reduction, $P=0.05$ for the interaction between gender and treatment). These findings reinforce the importance of lipid lowering in secondary prevention.

In the secondary-prevention LIPID trial, 9,014 men and women (31 to 75 years of age) with a history of either MI or unstable angina were treated for a mean follow-up period of 6.1 years with either pravastatin 40 mg/d or placebo.[24] Baseline cholesterol concentrations were 155 to 271 mg/dL, thereby characterizing these patients as having cholesterol in the average range found in the general US population. The primary end point, CHD mortality, was reduced by 24% ($P<0.001$). Among secondary end points, risk for death from any cause was reduced 22% ($P<0.001$); risk for nonfatal MI or CHD death by 24% ($P<0.001$); risk for stroke by 19% ($P=0.048$); and coronary revascularizations by 20%

($P<0.001$). LIPID was the second statin trial to demonstrate a reduction in the all-cause mortality rate. Results of a 2-year extended follow-up of the LIPID trial confirm the long-term efficacy and safety of statin therapy.[25]

In the primary-prevention AFCAPS/TexCAPS, 6,605 men (aged 45 to 73) and women (aged 55 to 73) with mean TC of 221 mg/dL, mean LDL-C of 150 mg/dL, mean HDL-C of 36 mg/dL (men) and 40 mg/dL (women), and median TG of 158 mg/dL, were treated for a mean of 5.2 years with either lovastatin (Mevacor®; Chapter 5) or placebo.[26] Lovastatin treatment was initiated at 20 mg/d; if LDL-C remained above 110 mg/dL after 12 weeks of treatment, the dosage was titrated to 40 mg/d. Treatment reduced LDL-C by a mean of 25% vs baseline and increased HDL-C by 6% vs baseline. The risk for a first acute major coronary event (defined as a composite of fatal or nonfatal MI, unstable angina, and sudden cardiac death) was reduced by 37% ($P<0.001$). AFCAPS/TexCAPS was the first primary-prevention statin trial to report clinical benefit in patients with average TC and LDL-C and below-average HDL-C levels.

Benefit in Subgroups of Patients

High-risk Subgroups: Heart Protection Study

Despite the clinical benefits of lipid-lowering therapy in the five first-generation major statin trials, evidence in certain types of high-risk patients was limited. The 5-year HPS, which was conducted in 20,536 patients, assessed the long-term effects of treatment in high-risk subgroups that were underrepresented in previous lipid-lowering trials (eg, women, the elderly, those with below-average LDL-C concentrations for Western societies). To be eligible for the study, participants were required to have one of the following: coronary disease, diabetes, noncoronary occlusive arterial disease, or treated hypertension (men ≥65 years).[4] At baseline, the mean nonfasting TC concentration was 228 mg/dL, LDL-C was 131 mg/dL, HDL-C was 41 mg/dL, and TG 186 mg/dL. Patients considered by their physicians to be definite candidates for lipid-lowering

Table 1-1: Major Placebo-Controlled Clinical Trials of HMG-CoA Reductase Inhibitor Therapy

	Primary Prevention	
	WOSCOPS	**AFCAPS/TexCAPS**
N (% women)	6,595 (0)	6,605 (15)
Duration (years)	4.9	5.2
Intervention	pravastatin 40 mg/d	lovastatin 20 to 40 mg/d
Baseline lipids (mg/dL)		
TC	272	221
LDL-C	192	150
HDL-C	44	36 men; 40 women
TG	164	158
% Lipid changes, treatment vs placebo		
TC	-20	-19
LDL-C	-26	-27
HDL-C	+5	+5
TG	-12	-13
End points (% changes in risk), treatment vs placebo		
Nonfatal MI/CHD death	**-31**	-25
Fatal/nonfatal MI	—	-40
Acute major coronary events*	—	**-37**
Total mortality	-22	+4 (ns)
CHD mortality	-28	too few for analysis
Revascularizations	-37	-33
Stroke	-11 (ns)	—

HMG-CoA=3-hydroxy-3-methylglutaryl coenzyme A; WOSCOPS=West of Scotland Coronary Prevention Study; AFCAPS/TexCAPS=Air Force/Texas Coronary Atherosclerosis Prevention Study; 4S=Scandinavian Simvastatin Survival Study; CARE=Cholesterol and Recurrent Events; LIPID=Long-Term Intervention with Pravastatin in Ischemic Disease; TC=total cholesterol; LDL-C=low-density lipoprotein cholesterol; HDL-C=high-density

Secondary Prevention

4S	CARE	LIPID
4,444 (19)	4,159 (14)	9,014 (17)
5.4	5	6.1
simvastatin 20** to 40 mg/d	pravastatin 40 mg/d	pravastatin 40 mg/d
261	209	218
188	139	150
46	39	36
133	155	140
-26	-20	-18
-36	-28	-25
+7	+5	+5
-17	-14	-11
-34	**-24**	-24
—	-25	-29
—	—	—
-30	-9 (ns)	-22
-42	-20	**-24**
-37	-27	-20
-30***	-31	-19

lipoprotein cholesterol; TG=triglyceride; MI=myocardial infarction; CHD=coronary heart disease; ns=nonsignificant. **bold**=study's primary end point; —=not reported

* Acute major coronary events are defined as the composite of fatal or nonfatal MI, unstable angina, and sudden cardiac death.

** Two patients were titrated down to 10 mg/d.

*** Includes stroke and transient ischemic attack.

Table 1-2: Heart Protection Study*

Study Design	
N (% women)	20,536 (25%)
Duration	5 years
Intervention (2 x 2 factorial design)	Simvastatin 40 mg/d; antioxidants: vitamin E 600 mg/d, vitamin C 250 mg/d, β-carotene 20 mg/d
Baseline Characteristics	
N with CHD	13,386 (65%)
N with	
Cerebrovascular disease	3,280** (1,820)†
Peripheral arterial disease	6,748** (2,701)†
Diabetes mellitus	5,963** (3,982)†
Baseline Lipids, mg/dL (mean)	
TC	228
LDL-C	131
HDL-C	41
TG	186
% Lipid Changes, Treatment vs Placebo (5-year change)	
TC	-13
LDL-C	-20
HDL-C	+2
TG	-9

End Points (% change in risk), Treatment vs Placebo

Primary (selected)	
Mortality:	
All-cause	-13[‡]
CHD	-18[§]
Secondary (selected)	
Major coronary events (nonfatal MI/CHD death)	-27[¶]
Major vascular events (major coronary events, stroke, any revascularization)	-24[¶]
Stroke (fatal/nonfatal)	-25[¶]

* High-risk patients with and without diagnosed CHD. For results in patients with diabetes, see *Lancet* 2003;361:2004-2016.

** Includes subjects with and without CHD (some with more than one of the three conditions).

† Subjects without CHD (some with more than one of the three conditions).

‡ P=0.0003

§ P=0.0005

¶ P<0.0001

CHD = coronary heart disease, HDL-C = high-density lipoprotein cholesterol, LDL-C = low-density lipoprotein cholesterol, MI = myocardial infarction, TC = total cholesterol, TG = triglycerides.

Data from *Lancet* 2002;360:7-22.

therapy were excluded. Subjects were randomly allocated to simvastatin 40 mg/d or placebo and, in a 2 x 2 factorial design, also received either an antioxidant regimen (600 mg vitamin E, 250 mg vitamin C, and 20 mg β-carotene daily) or placebo. The primary outcomes were all-cause mortality, CHD mortality, and mortality from all other causes; secondary end points were specific noncoronary causes of death, major coronary events (nonfatal MI or CHD death), major vascular events (major coronary events, stroke, coronary or noncoronary revascularization), and nonfatal or fatal strokes.

The HPS found a lower incidence of all-cause mortality with simvastatin vs placebo. This was based chiefly on a proportional 17% reduction in death from vascular causes ($P<0.0001$), which included a highly significant 18% reduction ($P=0.0005$) in coronary deaths and a 16% reduction in mortality from other vascular causes ($P=0.07$). There was no significant difference between simvastatin and placebo in nonvascular deaths. Simvastatin also produced proportional reductions of 38% and 27% in the rates of first nonfatal MI and nonfatal MI/CHD death ($P<0.0001$), respectively, and a 25% decrease in the incidence of stroke, which consisted mainly of 30% fewer ischemic strokes. No apparent difference between treatment groups in the risk for hemorrhagic stroke was observed. In all patients, simvastatin reduced the incidence of a major vascular event by 24% ($P<0.0001$). Similar risk reductions of approximately 20% occurred in patients whose LDL-C decreased by 37 mg/dL from baseline concentrations <116 mg/dL and ≥116 mg/dL. Older patients also benefited substantially from treatment. In the HPS, antioxidant therapy produced no significant reductions in cardiovascular or other major outcomes. The HPS results contribute to the rationale for a recent revision of the ATP III guidelines in which the cut point for considering lipid-lowering therapy in high-risk patients is now ≥100 mg/dL (reduced from ≥130 mg/dL), and an optional LDL-C goal of ≤70 mg/dL is favored in very high-risk patients (Table 1-2).

The HPS researchers concluded that 5 years of simvastatin therapy can prevent at least one major cardiovascular or cerebrovascular event in up to 100/1,000 high-risk patients, with longer-term therapy eventually producing greater benefit. This outcome is expected both in middle-aged and in elderly men and women without regard to initial cholesterol level.

Hypertensive Subgroups

In an observational study of 316,099 white men aged 35 to 57 years, strong graded relationships were found between CHD death and serum cholesterol levels >180 mg/dL, systolic blood pressure >110 mm Hg, and diastolic blood pressure >70 mm Hg.[6] The incidence of CHD death rose with increasing cholesterol levels in each quintile of blood pressure and with increasing blood pressure levels in each quintile of cholesterol, up to ≥245 mg/dL.[6] These positive associations, reflecting a multiplicative effect of cholesterol levels and blood pressure, persisted in all age groups over a 12-year follow-up period.[6]

ASCOT-LLA. For HPS subjects with treated hypertension (N=8,457), simvastatin reduced the risk for a major vascular event by approximately 20% vs placebo,[4] suggesting that statin therapy may provide additional protection in this subgroup. The recent ASCOT-LLA specifically tested the hypothesis that statin therapy decreases the incidence of cardiovascular events in hypertensive patients who are at moderate risk and are not conventionally considered dyslipidemic.[27] ASCOT is a large randomized trial (N=19,342) comparing two antihypertensive regimens for the prevention of CHD.[27] Participants in the double-blind ASCOT-LLA (N=10,305) were further randomized to atorvastatin (Lipitor®) 10 mg/day or placebo. Eligibility criteria included no history of CHD, treated or untreated hypertension according to specified criteria, ≥3 additional risk factors, a TC level ≤251 mg/dL, and no current statin or fibrate treatment.[27] In this largely white and male population, the mean baseline LDL-C concentration was 131 mg/dL,[27] as in the HPS.

The patients in this study experienced similar blood pressure control with each antihypertensive regimen.[27] In addition, atorvastatin produced a highly significant 36% reduction (*P*=0.0005) in the relative risk for a primary end point event (nonfatal MI, including silent MI, and fatal CHD), corresponding to a 29% decrease in calculated LDL-C levels.[27] There were also significant reductions in the secondary end points of total cardiovascular events (21%), total coronary events (29%), the primary end point excluding silent MI (38%), and fatal/nonfatal stroke (27%).[27] Effects on the secondary end points of heart failure or cardiovascular mortality did not differ significantly from those for placebo.[27]

The ASCOT-LLA indicates that lipid lowering can improve cardiovascular risk reduction in patients with good blood pressure control who, despite other risk factors, are at moderate risk and not conventionally deemed dyslipidemic.[27] As in the HPS, this supports the value of global risk assessment and the importance of treating high risk, not just high cholesterol.[4,28] Because of the early emergence of a significant treatment effect with atorvastatin, ASCOT-LLA was terminated after 3 years rather than continued for the planned duration of 5 years.[27]

ALLHAT-LLT. At first glance, results of the recent Antihypertensive and Lipid-Lowering Treatment to Prevent Heart Attack Trial (ALLHAT-LLT) might seem contrary to those of the ASCOT-LLA. However, the ALLHAT-LLT results were influenced by compliance and methodologic factors that affect their interpretation.

In the ALLHAT-LLT, a subset of 10,355 men and women aged ≥55 years with well-controlled hypertension and at least one other CHD risk factor was assigned in a nonblinded manner to receive pravastatin 40 mg/day or usual care. Lipid eligibility criteria included an LDL-C level of 120 to 189 mg/dL for patients with no known CHD and 100 to 129 mg/dL for those with known CHD.[29] Although moderately hypercholesterolemic as a whole, the ALLHAT subjects included 2,613 patients (25%) with LDL-C levels <130 mg/dL at baseline. Approximately 14% of the population had a history of CHD, and there were substantial numbers of women (49%), African Americans (34%), patients with diabetes (35%), and the elderly (ie, 55% aged ≥65 years).[29]

After 6 years of therapy, calculated LDL-C levels decreased by 30.1% with pravastatin and by 16.2% with usual care, a differential of 13.9% (mean LDL-C difference of 17.2 mg/dL).[29] This contrasts markedly with the large statin trials, in which the placebo-controlled groups experienced little or no cholesterol reduction.[29] With pravastatin, there were nonsignificant reductions of 1% in total mortality, the primary end point, and 9% in CHD death plus nonfatal MI, a secondary end point. For the secondary end point of cause-specific mortality, there was a nonsignificant decrease of 1% in CHD death.[29] Given the modest differential in LDL-C levels between the pravastatin and usual-care groups, the nonsignificant relative reductions in fatal and nonfatal coronary events are understandable.[29] By the study's end, 26% of usual-care subjects were taking a statin. The open-label design undoubtedly contributed to this crossover rate by enabling the community-based investigators to prescribe statin therapy based on clinical judgment.[30] Furthermore, medication adherence had declined to 77% in the pravastatin group. As a result, the ALLHAT-LLT intention-to-treat analysis seems to be based on approximately half of pravastatin-allocated patients actually taking the study medication (ie, 77% minus 26%).

The ALLHAT-LLT was estimated to provide 84% power to detect a 20% reduction in mortality, of which CHD death was a component.[29] In retrospect, this anticipated relative risk reduction appears too optimistic.[30] An analysis of the ALLHAT-LLT suggests that improved adherence and fewer crossovers could have led to a more robust LDL-C differential between the pravastatin and control groups, thereby producing cardiovascular risk reductions comparable to those in the other large statin trials.[29,30] Consequently, our efforts to reduce cardiovascular risk must include promoting patient compliance.

Acute Coronary Syndrome Subgroups

MIRACL. In the 16-week Myocardial Ischemia Reduction with Aggressive Cholesterol Lowering (MIRACL) trial (N=3,086), patients were randomized to atorvastatin 80 mg/

day or placebo within a mean of 63 hours after hospitalization for an acute coronary syndrome (ACS).[31] At the end of 16 weeks, there was a 16% reduction in risk (*P*=0.048) for a primary end-point event (death, nonfatal acute MI, cardiac arrest with resuscitation, or recurrent symptomatic myocardial ischemia requiring emergency hospitalization) with statin therapy. This was primarily the result of a 26% drop in recurrent symptomatic myocardial ischemia requiring hospitalization (*P*=0.02). From a baseline of 124 mg/dL in each group, LDL-C levels decreased to a mean of 72 mg/dL with atorvastatin and increased to 135 mg/dL with placebo. The marginally significant *P* value for a primary end-point event may be a consequence of the sample size, which was based on an expected 25% to 30% reduction in risk; therefore, a larger trial may produce a result of greater significance. However, the MIRACL results suggest that atorvastatin 80 mg/dL initiated during the acute phase of unstable angina or non-Q-wave MI can reduce the risk for an early recurrent cardiac event, primarily recurrent symptomatic ischemia.[31]

PROVE-IT. In the PROVE-IT study (N=4,162), which compared the effects of atorvastatin 80 mg/day with those of pravastatin 40 mg/day in recently hospitalized ACS patients, intensive acute-phase intervention reduced coronary risk both in the short term and over a more extended period of time.[32] After 2 years of treatment, there was a 16% reduction in the hazard ratio for cardiovascular events, the primary end point, with high-dose versus standard-dose therapy (*P*=0.005).[32] The benefit of high-dose therapy began to emerge at 30 days after randomization (a median of 7 days following onset of the inclusion event) and was consistent for the duration of the trial. Levels of LDL-C, which were 106 mg/dL in each treatment group at randomization, decreased to a median of 62 mg/dL with atorvastatin and 95 mg/dL with pravastatin. A significant between-group difference in C-reactive protein (CRP), a marker of inflammation, was also observed. For components of the primary end point, there were significant decreases in revascularizations (14%, *P*=0.04) and recurrent unstable angina (29%, $P=0.02$), as well as nonsignificant reductions in all-cause mortality (28%) and myocardial infarction (13%).[32] However, the incidence of hepatic adverse effects was significantly higher with aggressive treatment (3.3% vs 1.1%). In clinical practice, caution is needed because patients generally have more coexisting conditions than did the study participants and, therefore, may be less able to tolerate a high-dose regimen.[32]

A-to-Z. The Aggrastat-to-Zocor (A-to-Z) trial, also conducted in ACS patients, consisted of two sequential phases: an 'A' phase comparing open-label enoxaparin vs unfractionated heparin, and a double-blind 'Z' phase comparing two statin regimens in 4,500 stabilized subjects following either a non-ST-elevation ACS or an ST-elevation MI. Eligibility criteria included a TC level ≤250 mg/dL and at least one of several other high-risk characteristics. The primary end point comprised CVD death, nonfatal MI, readmission for ACS, and stroke.[33]

In the 'Z' phase, subjects were randomized to a more intensive (simvastatin 40 mg/d for 30 days, 80 mg/d thereafter) or less intensive (placebo for 4 months, simvastatin 20 mg/d thereafter) regimen. After 8 months, median LDL-C levels were 63 mg/dL and 77 mg/dL, respectively. This corresponded to a nonsignificant 11% reduction in risk for a primary end-point event. A number of explanations have been suggested for the lack of a significant outcome, including the possibility that statins differ in their nonlipid effects.

In patients with chronic atherosclerosis, 1 to 2 years are needed for benefit to emerge with statin therapy vs placebo. Although MIRACL and PROVE-IT imply that intensive treatment can prevent early recurrence of cardiac events following an ACS, it is unclear whether the benefit was attributable to lipid lowering or to plaque stabilization, or whether between-statin differences played a role.[32] In A-to-Z, markedly lower LDL-C levels with more intensive therapy at 30 days did not produce a corresponding reduction in cardiovascular events. During this initial period, CRP levels did not differ between intervention groups (although significant differences emerged at 4 and 8 months). Considered together with PROVE-IT and MIRACL, the results of A-to-Z suggest that the efficacy of statin therapy in ACS may be largely based on anti-inflammatory mechanisms, and that individual statins may differ in this regard.[34] Moreover, other types of lipid-lowering drugs may not exert anti-inflammatory effects; therefore, their efficacy in post-ACS patients cannot be assumed.[34]

Timing of treatment may play a role as well. For example, statin therapy at 80 mg/d was initiated within 7 days post ACS in PROVE-IT, and after 30 days in A-to-Z. In PROVE-IT, the between-group difference in CRP levels at 30 days (see above) suggests that immediate high-dose therapy may have produced an anti-inflammatory effect during the period of greatest instability, thus contributing to more rapid clinical benefit.[32,33] On the other hand, randomization at a median of 7 days after the index event meant that 69% of PROVE-IT subjects had undergone revascularization before being randomized, which may have stabilized the culprit lesion and acute thrombotic process. In A-to-Z, statin therapy was initiated 3 to 4 days earlier, and subjects were less likely to have undergone revascularization.[33]

The composition of the primary end point may also have affected the outcome. Although between-group differences in LDL-C were almost identical in MIRACL and A-to-Z at 4 months (63 mg/dL and 62 mg/dL, respectively), MIRACL subjects experienced a 16% reduction in risk, compared with no 4-month benefit in A-to-Z (the period of placebo treatment). The significant outcome in MIRACL was driven by a decrease in risk for recurrent symptomatic ischemia, a component not included in the A-to-Z primary end point.

In A-to-Z, nine subjects receiving simvastatin 80 mg/dL developed myopathy (creatine kinase level >10X upper limit of normal), compared with no cases in the 20-mg/d and 40-mg/d groups. Of the nine, rhabdomyolysis (creatine kinase level >10,000 units/L) was diagnosed in three subjects. In MIRACL and PROVE-IT, no cases of rhabdomyolysis were reported.[31-33] Although the incidence of rhabdomyolysis in A-to-Z may be attributed to chance, it is also possible that this indicates an adverse effect of high-dose simvastatin that is less likely to occur with certain other statins at the same dose.

GRACE. Recently, the Global Registry of Acute Coronary Events (GRACE) study reported that patients who experienced an ACS despite being on statin therapy appeared to have less severe presentation, fewer complications, and a lower incidence of mortality than did those who had never taken a statin. However, this large observational study (N=19,537), which has several limitations, can only be considered hypothesis generating.[35]

Safety of Statin Therapy

Because statins inhibit HMG-CoA reductase, a major liver enzyme, safety concerns have understandably focused on liver toxicity. Skeletal muscle function is also a concern.[21] The latter received considerable public attention when cerivastatin was withdrawn from the market because of an increased incidence of rhabdomyolysis.[36]

One objective of the PPP Project was to evaluate potential safety issues associated with long-term therapy by combining data from three randomized double-blind placebo-controlled trials of pravastatin 40 mg/d: WOSCOPS, CARE, and LIPID. Based on >112,000 person-years of experience, the results of this pooled analysis do not substantiate the concerns about muscle and liver toxicity.[21] No cases of myopathy (ie, muscle ache or weakness together with creatine kinase levels >10 ULN) were reported with pravastatin or placebo. There was a similar incidence of myalgia and/or myositis in both treatment groups, with no differences between older and younger subjects. Furthermore, serious hepatobiliary adverse events, including the risk of exacerbating a preexisting liver function abnormality, were similar with the study drug and placebo.

Overall, fewer deaths occurred in pravastatin patients, largely because of a reduction in cardiovascular mortality. The incidence of noncardiovascular adverse events, including death, was no greater with pravastatin than with placebo. Even after adjustment for potential confounders, including age, diabetes, smoking status, and a serious cardiovascular adverse event, there was a greater likelihood of continuing active therapy compared with placebo.

Some data have shown an inverse relationship between cholesterol levels and hemorrhagic stroke,[37,38] and it has been suggested that lipid-lowering therapy may produce a similar effect. However, the PPP Project does not support this concern.[22] An analysis of the CARE and LIPID database revealed a 22% lower incidence of stroke with pravastatin vs placebo (P=0.01). Although this was primarily the result of a reduction in nonhemorrhagic events, there was no between-group difference in the risk for a hemorrhagic stroke. According to these secondary-prevention data, an estimated 588 patients would require treatment per year to prevent one stroke. In WOSCOPS, a primary-prevention trial, stroke rates were lower than in the CARE/LIPID studies, but there was no benefit attributable to pravastatin therapy. Risk reduction was not significant, and 3,333 patients would have to be treated annually to avert one stroke.

The lack of a relationship between lipid lowering and the incidence of hemorrhagic stroke was also found in the HPS and in the MIRACL trial.[3,31,39,40] In the HPS, statin therapy significantly reduced the risk for stroke by 25% (P<0.0001) irrespective of age, sex, or lipid levels at baseline. This was primarily the result of a 30% decrease in the incidence of ischemic stroke (P<0.0001). There was no significant difference in hemorrhagic stroke with statin therapy vs placebo, even among participants whose LDL-C levels decreased from <116 mg/dL at baseline to an average of 70 mg/dL.[41] Similarly, results of a MIRACL substudy indicate that atorvastatin significantly lowered the overall incidence of stroke, a secondary end point, by 50%, primarily by reducing the incidence of thrombotic/embolic events.[39] There were only three hemorrhagic strokes, all with placebo, which suggests that intensive cholesterol lowering does not increase the risk for intracranial hemorrhage in ACS patients. However, the absolute number of events in this 16-week study was small, and these results need to be confirmed in future randomized trials. In PROVE-IT, stroke was a component of the primary end point. When examined individually, the incidence of stroke was infrequent and rates did not differ significantly between treatment groups. However, the type of stroke is not reported.[32]

Further evidence concerning the effect of lipid-modifying therapy on the incidence of type-specific stroke will be provided by the Cholesterol Treatment Trialists' Collaboration, a large meta-analysis now in progress.[41,42]

Regression Trials: Vascular Benefit of Lipid Lowering

Historically, atherosclerosis was considered inevitably progressive. Beginning in the late 1980s, clinical trials assessing disease severity, as measured either by coronary angiography or by ultrasonography of the carotid arteries, demonstrated that aggressive lipid management could halt the progress of atherosclerosis and, in some cases, even reverse existing disease.

These 'regression' studies have resulted in a number of important findings, one of which is that atherosclerotic disease continues to progress if left untreated. Another interesting finding is that treatment of patients with evidence of CHD usually does not lead to substantial reductions in coronary stenosis, but, in general, decreases the rate of progression. In addition, lipid-modifying drugs may have nonlipid (ie, 'pleiotropic') effects that can prevent the rupture of vulnerable plaque.[43]

Despite relatively small changes in coronary blockage with lipid-regulating therapy, unexpectedly large reductions in CHD events may occur.[44] Evidence has shown that progression of CHD blockage carries a markedly increased risk for a CHD event. The Program on the Surgical Control of the Hyperlipidemias (POSCH), in which partial ileal bypass surgery was the lipid-regulating intervention, demonstrates this point.[45] In POSCH, 838 men and women, 30 to 64 years old, with prior MI and hypercholesterolemia were randomized either to surgical intervention plus diet or to the diet control group. Mean follow-up was 9.7 years. The results of POSCH confirmed that aggressive lipid management could inhibit the progression of atherosclerotic disease. The lengthy follow-up provided an opportunity to demonstrate that angiographic changes predict future events, and, 5 years posttrial, significant reductions in overall and CHD mortality were observed.

Two trials have examined the effect of cholesterol lowering on vascular disease progression in patients with normal to mildly elevated baseline LDL-C concentrations. The first, the Harvard Atherosclerosis Reversibility Project (HARP), reported no angiographic benefit of treatment with pravastatin in this population.[46] However, in the Lipoprotein and Coronary Atherosclerosis Study (LCAS), disease progression was slowed in patients without severely elevated LDL-C and with angiographic evidence of CHD.[47] In LCAS, 429 men and women were randomized to treatment with either fluvastatin (Lescol®), 20 mg b.i.d., or placebo. Trial length was approximately 2.5 years. The primary end point was the change in minimum lumen diameter (ΔMLD) of qualifying lesions. At the end of follow-up, ΔMLD was –0.028 mm in fluvastatin-treated patients vs –0.100 mm in the control group (P<0.01), demonstrating slowed progression of atherosclerosis.

The recently completed Reversal of Atherosclerosis with Aggressive Lipid Lowering (REVERSAL) trial compared the effect of atorvastatin 80 mg/d vs pravastatin 40 mg/d on vascular end points, as evaluated by intravascular ultrasound (IVUS), in approximately 600 patients who underwent coronary angiography for a clinical indication and were found to have luminal narrowing of 20% to 50% in the target segment.[48] Subjects were also required to have an LDL-C level between 125 mg/dL and 210 mg/dL following a 4- to 10-week washout period of lipid-lowering medications. The primary end point was the percentage change in atheroma volume (between-group comparison). After 18 months, there was a nonsignificant negative change of -0.4% from baseline (median value) with atorvastatin (P=.98), signifying a halt in atherosclerosis progression, and a progression of 2.7% with pravastatin (P=.001). The between-group difference favoring intensive treatment was significant (P=.02). For the secondary end points of nominal change in total atheroma volume and change in percentage atheroma volume (median values), the study also found no progression with intensive treatment vs progression with standard treatment. For both secondary end points, the between-group comparisons were significant (P=.02 and P<.001, respectively). A correlation between vascular outcomes and clinical events remains to be established.

Refining Treatment Guidelines

Physician adherence to current treatment guidelines is essential in order to reduce the risk for CHD. However, a number of studies have documented inadequate adherence, including undertreatment of patients. For example, one cross-sectional analysis of medical records found that only 37% of 622 treatment-eligible secondary-prevention patients were receiving lipid-lowering therapy when admitted to the hospital for an acute MI.[49] Furthermore, only 11% of the patients with elevated cholesterol levels were prescribed a lipid-lowering drug on discharge. Multivariate analysis revealed advanced age (≥74 years) to be a significant risk factor for undertreatment. In the Lipid Treatment Assessment Project (L-TAP), a survey of 4,888 patients treated by primary care physicians, only 38.4% of patients achieved their NCEP target goal for LDL-C.[50] The success rate was lowest in CHD patients (18%). Those with two or more risk factors had a success rate of 37%, while 68% of patients at lowest risk achieved the target LDL-C level. Additional information about physician and patient compliance can be found in Chapters 3 and 4.

In addition to compliance with existing recommendations, periodic reevaluation based on the implications of recent clinical trials is essential. Accordingly, the ATP III has issued a 2004 update to its 2001 guidelines.[3] This update, which gives further support to the treatment of high risk, rather than just high cholesterol, is summarized in Chapter 3. Following is an examination of some major issues to be considered in future guideline revisions.

LDL-C: How Low to Go?

The 2001 ATP III guidelines recommend a target LDL-C value of <100 mg/dL in patients with established CHD or CHD risk equivalents (risk category 1). For patients in risk category 2 (≥2 risk factors and 10-year risk ≤20%), the LDL-C goal is <130 mg/dL, and for those in risk category 3 (0 to 1 risk factor and 10-year risk <10%), the LDL-C target is <160 mg/dL. In the Post-Coronary Artery Bypass Graft (Post-CABG) trial, aggressive lovastatin therapy to reach an LDL-C target <100 mg/dL yielded greater angiographic benefit than did a more moderate strategy,[51] and AFCAPS/TexCAPS reported a 37% reduction in the incidence of acute major coronary events in primary-prevention patients treated with lovastatin to an LDL-C goal below that recommended by the second NCEP Adult Treatment Panel.[26] More recently, the HPS found that decreasing LDL-C levels by 37 mg/dL reduced risk by approximately 20% in high-risk patients with initial lipid concentrations <116 mg/dL and ≥116 mg/dL.[4] Based largely on the HPS results, the ATP III update states that an LDL-C goal of <70 mg/dL is a therapeutic option in very high-risk patients.[3]

The Role of Triglycerides and HDL-C

Based on meta-analyses of prospective studies, the 2001 ATP III guidelines identify TG as an independent CHD risk factor. Once LDL-C goals are reached, non-HDL-C (which includes VLDL-C, the most readily available measure of TG-rich remnant lipoproteins) is considered a secondary therapeutic target in patients with TG levels of 200 to 499 mg/dL. Non-HDL-C levels can be reduced either by intensifying therapy with an LDL-C-lowering drug or by the cautious addition of fibrates or nicotinic acid to LDL-C-lowering therapy.

Once the LDL-C goal has been achieved, HDL-C becomes a therapeutic target only in high-risk patients with low HDL-C levels and TG levels <200 mg/dL (isolated low HDL-C). Two reasons are given: insufficient evidence to support a numeric goal, and a lack of available agents with a robust ability to raise HDL-C levels. The 2004 ATP III update is less restrictive, allowing for consideration of a fibrate or nicotinic acid in combination with an LDL-lowering drug in high-risk persons with TG elevations or low HDL-C.[3] This suggests that early therapy to lower TG or raise HDL-C levels is an option in the high-risk patient (see Chapter 5, Fibric Acid Derivatives, for precautions with combination therapy).

The interpretation of data regarding TG and HDL-C as coronary disease risk factors is complicated by a number of considerations. For example, in observational studies, the independent significance of TG as a predictor of coronary risk is weakened after multivariate adjustment for HDL-C, plasma glucose level, diabetes mellitus, and body mass index (BMI).[52,53] This suggests a complex interaction between TG and other metabolic risk factors (see The Metabolic Syndrome, below). In addition, limited clinical trial data imply that reducing coronary risk by lowering TG levels may depend on initial TG concentrations or on other lipid or nonlipid characteristics.[54,55]

Epidemiologic studies have shown that HDL-C is generally an independent predictor of CHD risk,[52,56-58] and that a 1-mg/dL increment is associated with an event reduction of 2% to 3%.[56-59] In the Veterans Affairs High-Density Lipoprotein Intervention Trial (VA-HIT), which enrolled diabetic and nondiabetic men with low HDL-C (<40 mg/dL) as their primary lipid abnormality, gemfibrozil vs placebo produced a 22% reduction in fatal/nonfatal CHD events.[60] This was associated with a 6% increase in HDL-C after 1 year (the between-group difference in LDL-C never differed significantly). Although on-treatment HDL-C was the only lipid to correlate significantly with a decrease in CHD after multivariate analysis, lipid levels achieved with therapy accounted for just 23% of the clinical benefit.[61] Therefore, other potentially favorable metabolic changes may contribute to the therapeutic effect of gemfibrozil. Moreover, the benefit of gemfibrozil was more dependent on the presence of insulin resistance than on the HDL-C level.[62] In the HHS, a study of gemfibrozil, there was a strong correlation between increased HDL-C levels and a reduced incidence of CHD[55]; interestingly, treatment benefit was most pronounced in subjects with a high BMI and other characteristics of the insulin resistance syndrome.[63]

Although many issues remain to be resolved regarding the link between high TG levels, low HDL-C levels, and coronary risk, the above data suggest that incrementally raising HDL-C levels may warrant further consideration as part of a CHD risk reduction strategy in a wider range of patients.

Emerging and Life-Habit Risk Factors

The increased CHD risk associated with elevated lipoprotein(a) [Lp(a)] levels also warrants closer investigation. In the ATP III guidelines, Lp(a) is identified as an emerging risk factor, as are homocysteine, prothrombotic/proinflammatory factors, impaired fasting glucose, and subclinical atherosclerotic disease. The guidelines also examine life-habit risk factors (ie, obesity, physical inactivity, atherogenic diet) and recommend TLC as essential to risk reduction. In the 2004 update, TLC is emphasized for high-risk patients with LDL-C levels ≥100 mg/dL or regardless of LDL-C level if lifestyle-related risk factors are present.

The Metabolic Syndrome

In many persons, life-habit, emerging, and major risk factors (addressed in Chapter 3) combine to create a condition called the metabolic syndrome. This condition is diagnosed when any three of the following are present: abdominal obesity, elevated TG, low HDL-C, hypertension, and impaired fasting glucose. Although the metabolic syndrome increases the risk for CHD at any LDL-C level, this complex association is not yet well characterized. Evidence suggests that the abnormalities of the metabolic syndrome may be manifestations of insulin resistance.[63] Moreover, because some epidemiologic data indicate that multivariate adjustment for HDL-C, glucose levels, diabetes, and BMI lessens the predictive significance of TG, it has been hypothesized that the relation between TG and CHD may be mediated by the metabolic consequences of hypertriglyceridemia.[64] In the view of some experts, TG-lowering therapy in diabetic, insulin-resistant, or other high-risk patients should be initiated when TG levels are ≥150 mg/dL, rather than ≥200 mg/dL (as advised by the ATP III guidelines).[65]

The Challenge

Continuing to refine treatment guidelines will be one of the most important challenges facing CHD researchers in the 21st century. The most pressing questions center on identifying the goals of therapy. Should LDL-C reduction continue to be the primary target for all patients? Should LDL-C levels be lowered as much as possible, or is there a threshold of benefit? What is the role of HDL-C in CHD risk and in risk-reduction therapy?

The important message of recent clinical trials is to treat high risk, rather than just high cholesterol. This requires a conceptual change in thinking: cholesterol levels considered 'average' in Western populations, even levels as low as 100 mg/dL, are not acceptable in high-risk patients. Furthermore, the terms 'primary' and 'secondary' prevention may represent a false distinction. As indicated in the ATP III guidelines, patients with CHD risk equivalents require a therapeutic approach as intensive as that in patients with prior CHD. The next step may be to enlarge our concept of high risk even further.

Atherosclerosis is a lifelong disease, and its presence in older patients is a function of LDL-C level and duration of exposure. The HPS and the Prospective Study of Pravastatin in the Elderly at Risk (PROSPER), another recent trial, have found substantial benefit of treatment in the elderly. This raises the question of whether lipid-lowering therapy initiated early in life can significantly reduce the burden of cardiovascular disease in later years. Clinical trials are needed to investigate this hypothesis. One related issue concerns the possible adverse effects of drug treatment over decades (eg, the effects of statins in women of childbearing age [see Appendix A, Case 5]).

Even in patients with CHD or CHD risk equivalents, it is important to measure lipid levels. In addition to TC and LDL-C, the lipid panel indicates HDL-C and TG levels. It is also the best available tool for determining whether a patient is compliant with lipid-lowering medication.

While most data now support reducing LDL-C levels as the primary goal of risk-reduction therapy, more research is needed to assess the role of other lipid subfractions and of nonlipid factors in the evolution of CHD. Studies must be conducted to develop drug and nondrug therapies that can modify these various risk factors to prevent CHD morbidity and mortality.

References

1. American Heart Association: *Heart Disease and Stroke Statistics—2004 Update*. Dallas, American Heart Association, 2004. Available at http://www.americanheart.org. Accessed August 19, 2004.

2. Expert Panel on Detection, Evaluation, and Treatment of High Blood Cholesterol in Adults: Executive Summary of the Third Report of the National Cholesterol Education Program (NCEP) Expert Panel on Detection, Evaluation, and Treatment of High Blood Cholesterol in Adults (Adult Treatment Panel III). *JAMA* 2001; 285:2486-2497.

3. Grundy SM, Cleeman JI, Merz CN, et al: Implications of recent clinical trials for the National Cholesterol Education Program Adult Treatment Panel III guidelines. *Circulation* 2004;110:227-239.

4. MRC/BHF Heart Protection Study of cholesterol lowering with simvastatin in 20,536 high-risk individuals: a randomised placebo-controlled trial. *Lancet* 2002;360:7-22.

5. Kannel WB: Cholesterol and risk of coronary heart disease and mortality in men. *Clin Chem* 1988;34(suppl 8B):B53-B59.

6. Neaton JD, Wentworth D: Serum cholesterol, blood pressure, cigarette smoking, and death from coronary heart disease. Overall findings and differences by age for 316,099 white men. Multiple Risk Factor Intervention Trial Research Group. *Arch Intern Med* 1992;152:56-64.

7. Pearson TA, LaCroix AZ, Mead LA, et al: The prediction of midlife coronary heart disease and hypertension in young adults: the Johns Hopkins multiple risk equations. *Am J Prev Med* 1990;6 (2 suppl):23-28.

8. Berenson GS, Srinivasan SR, Bao W, et al: Association between multiple cardiovascular risk factors and atherosclerosis in children and young adults. *N Engl J Med* 1998;338:1650-1656.

9. Raitakari OT, Juonala M, Kahonen M, et al: Cardiovascular risk factors in childhood and carotid artery intima-media thickness in adulthood: the Cardiovascular Risk in Young Finns Study. *JAMA* 2003;290:2277-2283.

10. McGill HC Jr, McMahan CA, Malcolm GT, et al: Effects of serum lipoproteins and smoking on atherosclerosis in young men and women. The PDAY Research Group. Pathobiological Determinants of Atherosclerosis in Youth. *Arterioscler Thromb Vasc Biol* 1997;17:95-106.

11. Strong JP, Malcolm GT, McMahan CA, et al: Prevalence and extent of atherosclerosis in adolescents and young adults: implications for prevention from the Pathobiological Determinants of Atherosclerosis in Youth study. *JAMA* 1999;281:727-735.

12. The Lipid Research Clinics Coronary Primary Prevention Trial results. I: Reduction in incidence of coronary heart disease. *JAMA* 1984;251:351-364.

13. The Lipid Research Clinics Coronary Primary Prevention Trial results. II: The relationship of reduction in incidence of coronary heart disease to cholesterol lowering. *JAMA* 1984;251:365-374.

14. Frick MH, Elo O, Haapa K, et al: Helsinki Heart Study: primary-prevention trial with gemfibrozil in middle-aged men with dyslipidemia. Safety of treatment, changes in risk factors, and incidence of coronary heart disease. *N Engl J Med* 1987;317:1237-1245.

15. Manninen V, Tenkanen L, Koskinen P, et al: Joint effects of serum triglyceride and LDL cholesterol and HDL cholesterol concentrations on coronary heart disease risk in the Helsinki Heart Study. Implications for treatment. *Circulation* 1992;85:37-45.

16. Canner PL, Berge KG, Wenger NK, et al: Fifteen year mortality in Coronary Drug Project patients: long-term benefit with niacin. *J Am Coll Cardiol* 1986;8:1245-1255.

17. Carlson LA, Rosenhamer G: Reduction of mortality in the Stockholm Ischaemic Heart Disease Secondary Prevention Study by combined treatment with clofibrate and nicotinic acid. *Acta Med Scand* 1988;223:405-418.

18. Randomised trial of cholesterol lowering in 4444 patients with coronary heart disease: the Scandinavian Simvastatin Survival Study (4S). *Lancet* 1994;344:1383-1389.

19. Baseline serum cholesterol and treatment effect in the Scandinavian Simvastatin Survival Study (4S). *Lancet* 1995;345:1274-1275.

20. Shepherd J, Cobbe SM, Ford I, et al: Prevention of coronary heart disease with pravastatin in men with hypercholesterolemia. West of Scotland Coronary Prevention Study Group. *N Engl J Med* 1995;333:1301-1307.

21. Pfeffer MA, Keech A, Sacks FM, et al: Safety and tolerability of pravastatin in long-term clinical trials. Prospective Pravastatin Pooling (PPP) Project. *Circulation* 2002;105:2341-2346.

22. Byington RP, Davis BR, Plehn JF, et al: Reduction of stroke events with pravastatin: the Prospective Pravastatin Pooling (PPP) Project. *Circulation* 2001;103:387-392.

23. Sacks FM, Pfeffer MA, Moye LA, et al: The effect of pravastatin on coronary events after myocardial infarction in patients with average cholesterol levels. Cholesterol and Recurrent Events Trial investigators. *N Engl J Med* 1996;335:1001-1009.

24. Prevention of cardiovascular events and death with pravastatin in patients with coronary heart disease and a broad range of initial cholesterol levels. The Long-Term Intervention with Pravastatin in Ischemic Disease (LIPID) Study Group. *N Engl J Med* 1998;339:1349-1357.

25. Long-term effectiveness and safety of pravastatin in 9014 patients with coronary heart disease and average cholesterol concentrations: the LIPID trial follow-up. *Lancet* 2002;359:1379-1387.

26. Downs JR, Clearfield M, Weis S, et al: Primary prevention of acute coronary events with lovastatin in men and women with average cholesterol levels: Results of AFCAPS/TexCAPS. Air Force/Texas Coronary Atherosclerosis Prevention Study. *JAMA* 1998;279:1615-1622.

27. Sever PS, Dahlöf B, Poulter NR, et al: Prevention of coronary and stroke events with atorvastatin in hypertensive patients who have average or lower-than-average cholesterol concentrations, in the Anglo-Scandinavian Cardiac Outcomes Trial—Lipid Lowering Arm (ASCOT-LLA): a multicentre randomised controlled trial. *Lancet* 2003;361:1149-1158.

28. Lindholm LH, Samuelsson O: What are the odds at ASCOT today? [commentary] *Lancet* 2003;361:1144-1145.

29. Major outcomes in moderately hypercholesterolemic, hypertensive patients randomized to pravastatin vs usual care. The Antihypertensive and Lipid-Lowering Treatment to Prevent Heart Attack Trial (ALLHAT-LLT). *JAMA* 2002;288:2998-3007.

30. Pasternak RC: The ALLHAT Lipid Lowering Trial—less is less. *JAMA* 2002;288:3042-3044.

31. Schwartz GG, Olsson AG, Ezekowitz MD, et al: Effects of atorvastatin on early recurrent ischemic events in acute coronary syndromes: the MIRACL study: a randomized controlled trial. *JAMA* 2001;285:1711-1718.

32. Cannon CP, Braunwald E, McCabe CH, et al: Intensive versus moderate lipid lowering with statins after acute coronary syndromes. *N Engl J Med* 2004;350:1495-1504.

33. de Lemos JA, Blazing MA, Wiviott SD, et al: Early intensive vs a delayed conservative simvastatin strategy in patients with acute coronary syndromes: phase Z of the A to Z trial. *JAMA* 2004;292:1307-1316.

34. Nissen SE: High-dose statins in acute coronary syndromes (editorial). *JAMA* 2004;292:1365-1367.

35. Spencer FA, Allegrone J, Goldberg RJ, et al: Association of statin therapy with outcomes of acute coronary syndromes: the GRACE study. *Ann Intern Med* 2004;140:857-866.

36. Fischer C, Wolfe SM, Sasich L, et al: Petition to the FDA to issue strong warnings about the potential for certain cholesterol-lowering drugs to cause potentially life-threatening muscle damage [letter]. HRG Publication #1588. August 20, 2001. Available at http://www.citizen.org/publications/release.cfm?ID=7051. Accessed October 15, 2002.

37. Yano K, Reed DM, MacLean CJ: Serum cholesterol and hemorrhagic stroke in the Honolulu Heart Program. *Stroke* 1989;20:1460-1465.

38. Iso H, Jacobs DR Jr, Wentworth D, et al: Serum cholesterol levels and six-year mortality from stroke in 350,977 men screened for the Multiple Risk Factor Intervention Trial. *N Engl J Med* 1989; 320:904-910.

39. Waters DD, Schwartz GG, Olsson AG, et al: Effects of atorvastatin on stroke in patients with unstable angina or non-Q-wave myocardial infarction: a Myocardial Ischemia Reduction with Aggressive Cholesterol Lowering (MIRACL) substudy. *Circulation* 2002;106:1690-1695.

40. Gotto AM Jr, Farmer JA: Reducing the risk for stroke in patients with myocardial infarction: a Myocardial Ischemia Reduction with Aggressive Cholesterol Lowering (MIRACL) substudy. *Circulation* 2002;106:1595-1598.

41. Collins R, Armitage J, Parish S, et al: Effects of cholesterol-lowering with simvastatin on stroke and other major vascular events in 20 536 people with cerebrovascular disease or other high-risk conditions. *Lancet* 2004;363:757-767.

42. Protocol for a prospective collaborative overview of all current and planned randomized trials of cholesterol treatment regimens. Cholesterol Treatment Trialists' (CTT) Collaboration. *Am J Cardiol* 1995;75:1130-1134.

43. Gotto AM Jr, Farmer JA: Pleiotropic effects of statins: do they matter? *Curr Opin Lipidol* 2001;12:391-394.

44. Brown BG, Zhao XQ, Sacco DE, et al: Lipid lowering and plaque regression. New insights into prevention of plaque disruption and clinical events in coronary disease. *Circulation* 1993; 87:1781-1791.

45. Buchwald H, Varco RL, Boen JR, et al: Effective lipid modification by partial ileal bypass reduced long-term coronary heart disease mortality and morbidity: five-year posttrial follow-up report from the POSCH. Program on the Surgical Control of the Hyperlipidemias. *Arch Intern Med* 1998;158:1253-1261.

46. Sacks FM, Pasternak RC, Gibson CM, et al: Effect on coronary atherosclerosis of decrease in plasma cholesterol concentrations in normocholesterolaemic patients. Harvard Atherosclerosis Reversibility Project (HARP). *Lancet* 1994;344:1182-1186.

47. Herd JA, Ballantyne CM, Farmer JA, et al: Effects of fluvastatin on coronary atherosclerosis in patients with mild to moderate cholesterol elevations (Lipoprotein and Coronary Atherosclerosis Study [LCAS]). *Am J Cardiol* 1997;80:278-286.

48. Nissen, SE, Tuzcu EM, Schoenhagen P, et al: Effect of intensive compared with moderate lipid-lowering therapy on progression of coronary atherosclerosis. A randomized controlled trial. *JAMA* 2004;291:1071-1080.

49. Majumdar SR, Gurwitz JH, Soumerai SB: Undertreatment of hyperlipidemia in the secondary prevention of coronary artery disease. *J Gen Intern Med* 1999;14:711-717.

50. Pearson TA, Laurora I, Chu H, et al: The Lipid Treatment Assessment Project (L-TAP): a multicenter survey to evaluate the percentages of dyslipidemic patients receiving lipid-lowering therapy and achieving low-density lipoprotein cholesterol goals. *Arch Intern Med* 2000;160:459-467.

51. The effect of aggressive lowering of low-density lipoprotein cholesterol levels and low-dose anticoagulation on obstructive changes in saphenous-vein coronary-artery bypass grafts. The Post Coronary Artery Bypass Graft Trial Investigators. *N Engl J Med* 1997;336:153-162.

52. Assmann G, Schulte H: Relation of high-density lipoprotein cholesterol and triglycerides to incidence of atherosclerotic coronary artery disease (the PROCAM experience). Prospective Cardiovascular Munster study. *Am J Cardiol* 1992;70:733-737.

53. Criqui MH, Heiss G, Cohn R, et al: Plasma triglyceride level and mortality from coronary heart disease. *N Engl J Med* 1993; 328:1220-1225.

54. Secondary prevention by raising HDL cholesterol and reducing triglycerides in patients with coronary artery disease: the Bezafibrate Infarction Prevention (BIP) study. *Circulation* 2000; 102:21-27.

55. Manninen V, Elo MO, Frick MH, et al: Lipid alterations and decline in the incidence of coronary heart disease in the Helsinki Heart Study. *JAMA* 1988;260:641-651.

56. Gordon DJ, Rifkind BM: High-density lipoprotein—the clinical implications of recent studies. *N Engl J Med* 1989;321:1311-1316.

57. Gordon DJ, Probstfield JL, Garrison RJ, et al: High-density lipoprotein cholesterol and cardiovascular disease. Four prospective American studies. *Circulation* 1989;79:8-15.

58. Assmann G, Cullen P, Schulte H: The Munster Heart Study (PROCAM). Results of follow-up at 8 years. *Eur Heart J* 1998;19 (suppl A):A2-A11.

59. Boden WE: High-density lipoprotein cholesterol as an independent risk factor in cardiovascular disease: assessing the data from Framingham to the Veterans Affairs High-Density Lipoprotein Intervention Trial. *Am J Cardiol* 2000;86:19L-22L.

60. Rubins HB, Robins SJ, Collins D, et al: Gemfibrozil for the secondary prevention of coronary heart disease in men with low levels of high-density lipoprotein cholesterol. Veterans Affairs High-Density Lipoprotein Intervention Trial Study Group. *N Engl J Med* 1999;341:410-418.

61. Robins SJ, Collins D, Wittes JT, et al: Relation of gemfibrozil treatment and lipid levels with major coronary events: VA-HIT: a randomized controlled trial. *JAMA* 2001;285:1585-1591.

62. Tenkanen L, Manttari M, Manninen V: Some coronary risk factors related to the insulin resistance syndrome and treatment with gemfibrozil. Experience from the Helsinki Heart Study. *Circulation* 1995;92:1779-1785.

63. Robins SJ, Rubins HB, Faas FH, et al: Insulin resistance and cardiovascular events with low HDL cholesterol: the Veterans Affairs HDL Intervention Trial (VA-HIT). *Diabetes Care* 2003;26: 1513-1517.

64. Reaven G: Metabolic syndrome: pathophysiology and implications for management of cardiovascular disease. *Circulation* 2002;106:286-288.

65. Grundy SM, Vega GL: Two different views of the relationship of hypertriglyceridemia to coronary heart disease. Implications for treatment. *Arch Intern Med* 1992;152:28-34.

Fundamentals of Blood Lipid Metabolism and Concepts in Atherogenesis

The roles of the various blood lipoproteins in atherogenesis are becoming better understood. A lipoprotein is a complex macromolecule of lipid and protein in which the nonpolar lipid core is surrounded by an amphipathic polar monolayer of phospholipids, heads of free cholesterol, and apolipoproteins, with protruding hydroxide groups (Figure 2-1).[1,2] This arrangement allows for the transport of hydrophobic, insoluble lipids through aqueous plasma. Embedded in the surface monolayer of the lipoprotein, apolipoproteins (also called 'apoproteins') serve multiple functions in determining the metabolic fate of the lipoprotein, including activating lipolytic enzymes and serving as ligands in receptor-mediated processes. Lipoproteins, which differ in the content of their lipid core, in the proportion of the lipids within the core, and in the proteins found on their surface, are classified according to density and electrophoretic mobility (Table 2-1). The density of the lipoprotein reflects the ratio of apolipoprotein to lipid. Lipid disorders may alter the composition and structure of the lipoprotein. For example, hypertriglyceridemic patients often produce low-density lipoprotein (LDL) particles that are small, dense, and have a relatively high protein-to-lipid ratio.

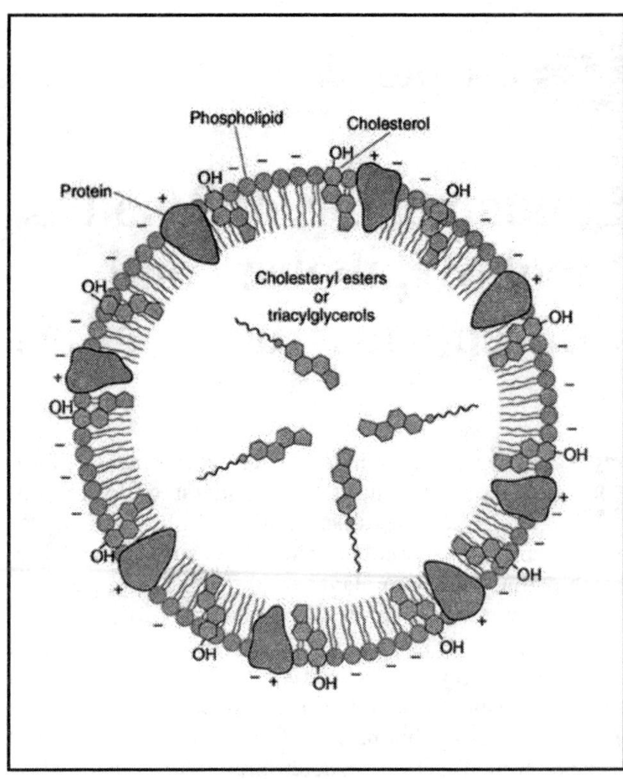

Figure 2-1: Structure of a lipoprotein. The core consists of nonpolar lipid molecules, which are hydrophobic. The protrusion of the polar hydroxide (OH) group through the surface monolayer makes the lipoprotein available to bind with the aqueous plasma. As a result, the lipoprotein is soluble in blood, which enables it to transport lipids through the bloodstream. The lipid content of the core varies with the class of particle. For example, chylomicrons and VLDL are rich in triglyceride, whereas LDL is rich in cholesteryl esters. VLDL = very-low-density lipoprotein; LDL = low-density lipoprotein.
From: Denniston KJ: *General, Organic, and Biochemistry,* 3rd ed. New York, NY, McGraw-Hill, 2000, p 516.

Lipid Synthesis

Lipoproteins contain one or more of the following lipids in varying combinations and proportions: esterified cholesterol, triglyceride (TG), and phospholipids (Table 2-1). This chapter begins its review of lipid metabolism by briefly describing the synthesis of fatty acids (a basic component of TG) and cholesterol (Figure 2-2).[1] Through covalent association, these simple lipids combine with carbon, hydrogen, or oxygen atoms to form complex lipids.[3]

Fatty Acids

Produced in various tissues, including hepatic, intestinal, and adipose, fatty acids are synthesized from glucose through a complex lipogenic pathway that is highly regulated by the hormones insulin, glucagon, and somatostatin. The availability of initial substrates and products of this pathway also influences lipogenesis. In the liver, fatty acids are incorporated into TG and secreted in very-low-density lipoprotein (VLDL) particles. In the intestine, TG is secreted in chylomicrons. Enhanced fatty acid synthesis in the liver may account, in part, for the hypertriglyceridemia associated with diabetes mellitus.

Cholesterol

The body uses free cholesterol, a simple lipid, for numerous functions, including cell membrane biogenesis, steroid synthesis, and the formation of bile acids. Cholesteryl esters, which are complex lipids derived from cholesterol, can be found in the lipid core of chylomicron remnants and intermediate-density, low-density, and high-density lipoproteins.

The human body can produce all the cholesterol it requires, although researchers estimate that about 20% to 40% of cholesterol is obtained through diet. Intake of exogenous cholesterol can down-regulate endogenous production to some degree. The liver is the primary producer of endogenous cholesterol and the primary processor of dietary cholesterol.

Table 2-1: Lipoprotein Classes and Associated Lipids and Apolipoproteins

Lipoprotein Class/ Density (g/mL)	Major Lipid Components*
Chylomicrons/≤0.95	Dietary triglycerides
Chylomicron remnants/0.96-1.006	Dietary cholesteryl esters
VLDL/<1.006	Endogenous triglycerides
IDL/1.006-1.019	Endogenous cholesteryl esters Endogenous triglycerides
LDL**/1.019-1.063	Endogenous cholesteryl esters
HDL_2/1.063-1.125	Cholesteryl esters, phospholipids
HDL_3/1.125-1.210	Phospholipids

*Chylomicrons, which contain dietary triglycerides, are assembled and secreted into the bloodstream by the intestines (ie, exogenous pathway of lipid metabolism). Endogenously produced triglycerides and cholesterol (contained in VLDL, IDL, LDL) are synthesized, secreted, and catabolized via the liver (ie, endogenous pathway). The synthesis and catabolism of HDL occur both in the intestines and in the liver.
**Lp(a), a lipoprotein subclass, consists of an LDL particle to which apolipoprotein(a) is bound. Apo B-100 is linked to apo(a) by a disulfide bond. The lipid core of Lp(a) consists mainly of cholesteryl esters.
VLDL = very-low-density lipoprotein; IDL = intermediate-density lipoprotein; LDL = low-density lipoprotein, HDL = high-density lipoprotein; Lp(a) = lipoprotein(a).

Cholesterol is derived in vivo from acetate via a mechanism with a rate-limiting step in which 3-hydroxy-3-methylglutaryl coenzyme A (HMG-CoA) is converted into mevalonic acid (mevalonate) by HMG-CoA reductase (Figure 2-3). This

Major Apolipoproteins	Electrophoretic Mobility
A-I, A-II, A-IV, B-48, C-I, C-II, C-III, E***	Origin
B-48, E***	Origin
B-100, C-I, C-II, C-III, E***	Preβ
B-100, C-I, C-II, C-III, E***	Broad-β
B-100	β
A-I, A-II, C-I, C-II, C-III, E	α
A-I, A-II, C-I, C-II, C-III	α

***The apo E isoforms are named E1, E2, E3, and E4, corresponding with the alleles ε1, ε2, ε3, and ε4. Apo E isoforms are inherited in a codominant fashion. Thus, eight phenotypes are possible: E1/E1, E1/E2, E1/E3, E2/E2, E2/E3, E2/E4, E3/E3, and E3/E4. These phenotypes have variable affinity for the LDL receptor. For example, the E3/E3 phenotype is associated with normal chylomicron remnant metabolism, but the E2/E2 phenotype is associated with elevated remnant levels, because the E2 isoform has less affinity for the LDL receptor.

Adapted from Gotto AM Jr, Pownall H: *Manual of Lipid Disorders*, 3rd ed. Philadelphia, Lippincott Williams & Wilkins, 2003, p 6.

step is inhibited by lipid-lowering drugs known as HMG-CoA reductase inhibitors, or statins. Such inhibition yields the significant decreases in endogenous cholesterol production and serum cholesterol levels that occur with these agents.

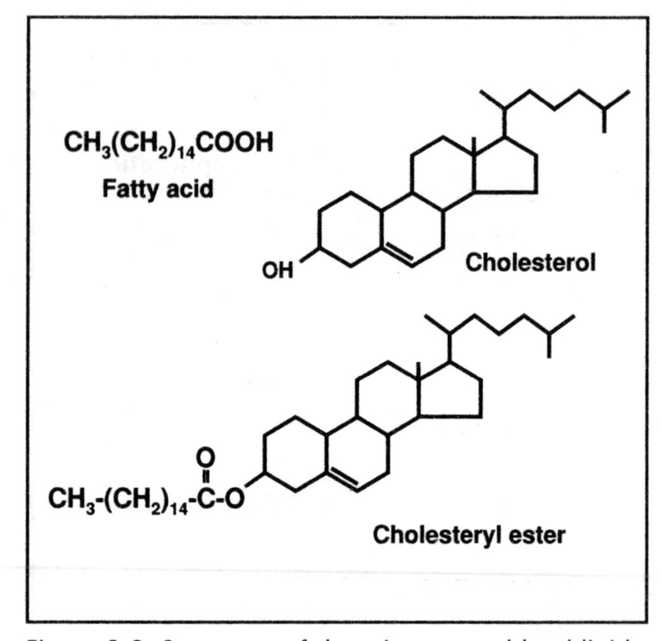

Figure 2-2: Structures of three important blood lipids. Fatty acids are esterified with a glycerol backbone to form triglyceride. Cholesterol is water insoluble and is esterified to the slightly soluble form of cholesteryl ester to facilitate transport in lipoproteins through the aqueous milieu.

The Lipoproteins

The major lipoproteins are chylomicrons, VLDL, intermediate-density lipoprotein (IDL), LDL, and high-density lipoprotein (HDL).[3] Two species of HDL have been identified: HDL_2 and HDL_3. Epidemiologic evidence also supports a role for lipoprotein(a) [Lp(a)] in coronary heart disease (CHD) risk. Lipoprotein(a) is a particle consisting of LDL bound to apolipoprotein(a) [apo(a)], a protein that may interfere with thrombolysis.[3]

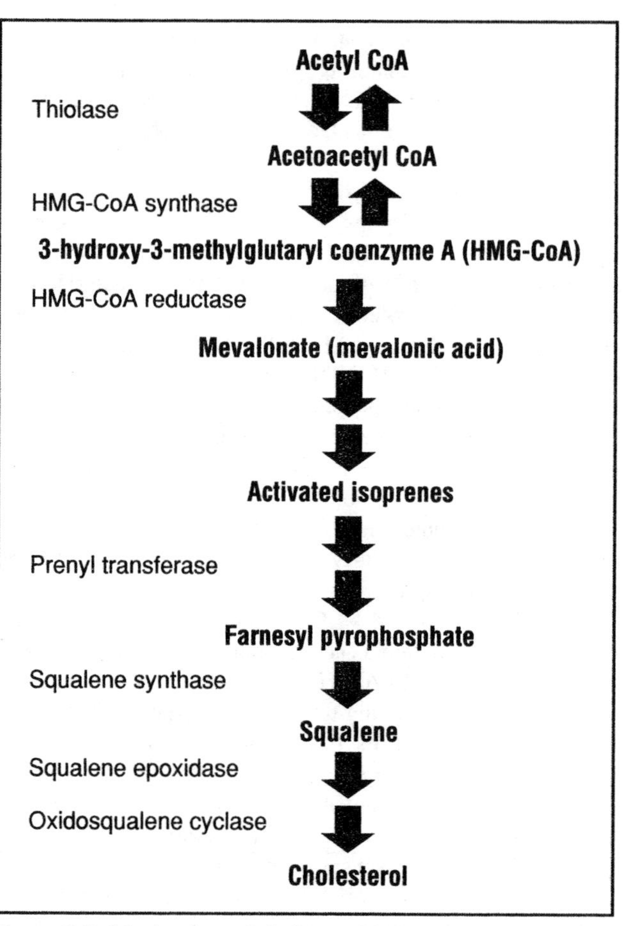

Figure 2-3: Mechanism of cholesterol biosynthesis. The statins interrupt cholesterol synthesis at the HMG-CoA reductase step.

The Apolipoproteins

The apolipoproteins are the chief structural components of the lipoproteins. Researchers originally speculated that apolipoproteins belonged to distinct families that were identified by an alphabetical nomenclature. Although this hypothesis has fallen out of favor, the naming system that it inspired remains. The Roman numeral suffix describes the order in which the apolipoproteins emerge from a chromatographic column.

Apolipoproteins A and B

The principal apolipoproteins of HDL are apolipoprotein (apo) A-I and apo A-II.[4] High-density lipoproteins belong to a heterogeneous family of particles that are characterized by density (ie, apolipoprotein/cholesterol ratio), apolipoprotein composition (ie, LpA-I or LpA-I:A-II), or electrophoretic mobility (ie, α or preβ).[5] HDL originates as lipid-poor complexes of LpA-I and phospholipids.[6] These discoidal particles, which have preβ electrophoretic mobility, are highly efficient acceptors of unesterified cholesterol from cells, the first step in reverse cholesterol transport (reviewed below).[5,6] They can be secreted intact from hepatocytes or enterocytes or generated extracellularly by the interaction of lipid-poor apo A-I with phospholipids either from cell membranes or from chylomicrons and VLDL, which shed excess surface constituents when their TG components undergo lipolysis.[6,7] Once formed, preβ LpA-I HDL particles undergo continuous, extensive remodeling to mature into spherical α-HDL. This process involves the acquisition of additional apolipoproteins, including apo A-II. Although the mechanisms for the formation of A-I/A-II HDL have become better understood in recent years, much is still unclear[4,6] (see Reverse Cholesterol Transport, this chapter).

Studies suggest that apo A-I and apo B (the principal apolipoprotein component of LDL-C) may be better than total cholesterol (TC), LDL-C, HDL-C, and TG as predictors of CHD.[8,9] In the Air Force/Texas Coronary Atherosclerosis Prevention Study (AFCAPS/TexCAPS) cohort, which was characterized by average LDL-C levels and below-average concentrations of HDL-C, apo A-I was the most consistent predictor of subsequent coronary events at baseline, at 1-year follow-up, and based on the change from baseline to follow-up.[10] Apo B was the most significant predictor of coronary risk at baseline and at 1-year follow-up. This significance was enhanced slightly by the incorporation of apo-A-I to form the apo B/apo A-I ratio, which was the best baseline indicator of risk.[10] In the Bogalusa Heart Study, low concentrations of HDL-C and apo A-I at puberty were among the factors that placed young white men at risk for CHD later in life.[11]

The apo B proteins are crucial to the production of chylomicrons and VLDL. Apolipoprotein B-48 is a truncated form of apo B-100; it is homologous with the first 2,151 amino acids of the amino terminal of apo B-100 and has 48% less mass than apo B-100. Apolipoprotein B-48 is coded for by intestinal cells, whereas hepatocytes code for apo B-100 synthesis. Therefore, lipids derived from exogenous sources are packaged in lipoproteins containing apo B-48 (chylomicrons), whereas endogenously produced lipids are packaged in particles containing apo B-100 (VLDL, IDL, LDL). Apolipoprotein B-100-containing particles are cleared through the LDL receptor, which also recognizes apo E, but not apo B-48. Microsomal transfer protein is required for the synthesis and secretion of apo-B-containing lipoproteins. Microsomal transfer protein is absent in the genetic disorder abetalipoproteinemia, which is characterized by hypocholesterolemia and other deleterious sequelae, and its regulation represents a promising target for future intervention. Apo B has been suggested as a more sensitive risk marker than TC or LDL-C because it more accurately reflects the presence of all atherogenic lipoproteins.[10]

Apolipoproteins C and E

The apo C proteins regulate VLDL catabolism. Apolipoprotein C-II activates lipoprotein lipase (LPL), a significant lipolytic enzyme. Apolipoprotein C-III appears to prevent the cellular uptake of apo B- and E-containing lipoproteins.

Apolipoprotein E is involved with chylomicron metabolism. It is polymorphic, multiallelic, and has a 20-fold greater affinity than apo B-100 for the cellular receptor that clears

LDL from the bloodstream. The genetic variants of apo E have different affinities for the LDL receptor; hence, apo E-mediated clearance of lipoproteins is somewhat determined by the phenotype (Table 2-1).

The LDL Receptor

In the ground-breaking, Nobel Prize-winning work of Brown and Goldstein, the severely elevated cholesterol associated with familial hypercholesterolemia, a genetic disorder, was attributed to a defect in the cellular receptor for LDL.[12] This LDL receptor is also called the B/E receptor because of its ability to recognize particles containing either apo B or apo E. Approximately 60% to 70% of LDL receptor activity occurs in the liver, although other body cells are also able to take up LDL. An important feature of the LDL receptor is its binding selectivity: as stated previously, the receptor recognizes apo E more readily than apo B-100, thus accounting for the more rapid clearance from plasma of lipoproteins containing apo E.

Exogenous Lipid Metabolism

Serum cholesterol is derived from two sources: the endogenous pathway (cholesterol biosynthesis) and the exogenous pathway (intestinal uptake of dietary and biliary cholesterol). Chylomicrons, which consist largely of triglycerides derived from dietary fat, are responsible for transporting exogenous (dietary) lipids through the bloodstream.

Synthesized in the mucosa of the small intestine, chylomicrons contain apolipoproteins A-I, A-II, and B-48. As chylomicrons circulate through the lymphatic system, they exchange apo A-I and apo A-II for apolipoproteins C and E from HDL. Through the action of cholesteryl ester transfer protein (CETP), chylomicrons also exchange TG for cholesteryl esters from HDL (Figure 2-4).

In the capillary beds of the arterial system, apo C-II activates LPL. Found in the capillary endothelium, LPL hydrolyzes the TG content of the chylomicron into free fatty

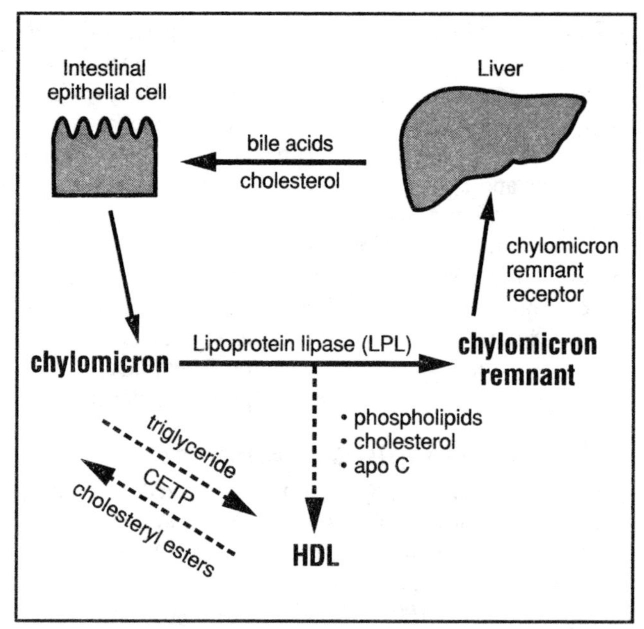

Figure 2-4: Pathway of exogenous (dietary) lipid metabolism. This simplified schema depicts the main steps that have been identified in this pathway. CETP=cholesteryl ester transfer protein; HDL=high-density lipoprotein.

acids (FFAs). Free fatty acids obtained in this fashion are either oxidized by muscle cells for energy, stored by adipose tissue, or returned to the liver for oxidation or for reesterification to TG used in VLDL synthesis.

The enzymatic action of LPL and the protein and lipid exchanges with HDL reduce the chylomicron to a smaller particle called a chylomicron remnant. Compared with their precursors, chylomicron remnants are smaller, more dense, and relatively depleted of TG and apo C, but retain apo B-48 and apo E, as well as cholesteryl esters from the intestinal lumen. Because remnant lipoproteins lose their TG content gradually, the blood contains a spectrum of particles at different stages of lipolysis and with varying proportions of TG and cholesterol.[13] Remnants are cleared from the circulation by uptake into hepatocytes through a chylomicron remnant receptor (possibly an LDL receptor-like protein) that recognizes apo E. Apolipoprotein C-III inhibits the uptake of apo E-containing remnant particles by the liver.

Chylomicron expression is closely linked to dietary intake of fat: increased fat consumption leads to increased chylomicron production. Chylomicrons are not found in the postabsorptive state (9- to 15-hour fast), and a high-carbohydrate, low-fat diet diminishes their formation. Under such conditions, TG transport is achieved through VLDL particles.

Endogenous Lipid Metabolism

Very-low-density lipoprotein is synthesized in the liver (Figure 2-5) either from FFAs obtained from chylomicron catabolism or from endogenously produced TG. These particles are smaller and more dense than chylomicrons. Apolipoproteins B-100, C-I, C-II, C-III, and E are associated with VLDL. A high-carbohydrate diet can increase VLDL-TG production.

Through the action of CETP, VLDL exchanges TG for cholesteryl esters from HDL. As with chylomicrons, LPL catalyzes the hydrolysis of TG in VLDL to FFAs that are either used by muscle or stored in adipose tissue. Triglyceride hydrolysis catalyzed by LPL reduces the size of the VLDL particle, resulting in IDL. Intermediate-density lipoprotein can either be taken up by the LDL receptor or be further reduced to LDL through the action of hepatic lipase (HL). Intermediate-density lipoprotein clearance is mediated by apo E.

Of the lipoproteins, only LDL contains just apo B-100. Low-density lipoprotein is cleared by both receptor-mediated and nonreceptor-mediated processes. Approximately two thirds of LDL is taken up through the LDL receptor. Peripheral cells can also take up LDL for use in membrane biogenesis and steroid synthesis. About 70% of plasma cho-

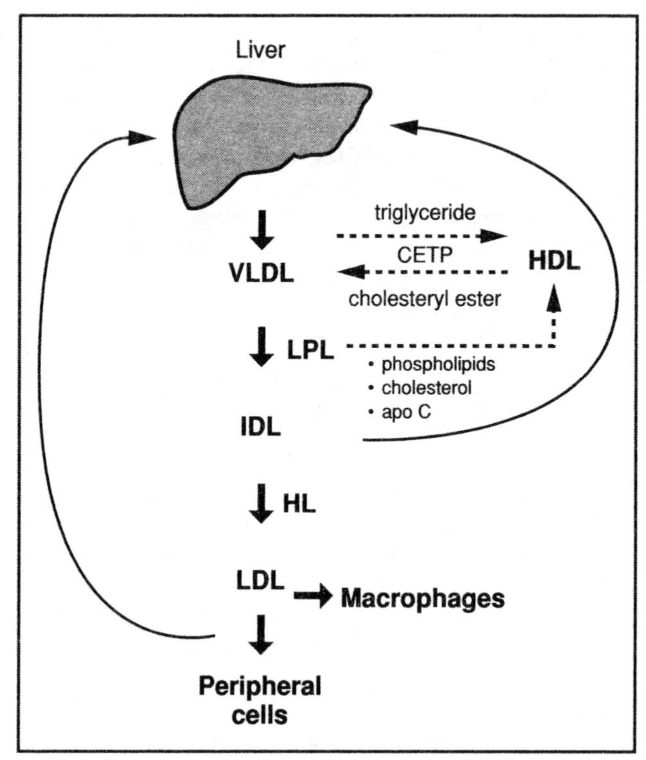

Figure 2-5: Pathway of endogenous lipid metabolism. This simplified schema depicts the main steps that have been identified in this pathway. CETP mediates the exchange of cholesteryl ester and triglycerides between HDL and apo B-containing lipoproteins (VLDL, IDL, LDL). CETP = cholesteryl ester transfer protein; LPL = lipoprotein lipase; HL = hepatic lipase; HDL = high-density lipoprotein; IDL = intermediate-density lipoprotein; LDL = low-density lipoprotein; VLDL = very-low-density lipoprotein.

lesterol is carried by LDL particles. The half-lives of VLDL and IDL are brief (approximately 12 hours), compared with LDL (2.5 to 3.5 days).

Modified LDL

Low-density lipoproteins may be modified through acylation, oxidation, or both. These modified lipoproteins are particularly important in our examination of atherogenesis. First, these molecules are cytotoxic and may damage the vascular endothelium, thus initiating the proliferative process of atherosclerosis. Second, these particles may aggregate in the intima of the vessel wall, where they are chemotactic to T cells and monocytes. Modified LDL has decreased affinity for the LDL receptor and may be cleared primarily through a scavenger receptor pathway. These scavenger receptors may explain the mode of transport of lipid into macrophages.

Small, Dense LDL

A derangement of LDL metabolism creates a species of LDL particle that has a higher protein-to-lipid ratio. Elevated concentrations of these small, dense LDL particles are associated both with increased CHD risk and with elevated TG and low HDL-C levels. A possible mechanism that would explain the link between small, dense LDL and hypertriglyceridemia involves the production of chylomicrons and VLDL particles that are particularly rich in TG. Catabolism of these TG-saturated VLDL particles produces LDL particles that contain a greater proportion of TG (normally about 10% of the particle). This makes the LDL a substrate for further hydrolysis by HL, resulting in decreased size and increased density. Small, dense LDL particles are believed to be more susceptible to oxidative modification. Hence, they are thought to be highly atherogenic. Metabolic disorders accompanied by hypertriglyceridemia, such as diabetes and the insulin resistance syndrome (or the metabolic syndrome), often produce this lipoprotein profile.

Lipoprotein(a)

Lipoprotein(a) is a particle comparable in size to LDL. It is assembled from an LDL particle and a large, hydrophilic glycoprotein called apolipoprotein(a) [apo(a)],[14] a plasminogen-like protein. Apolipoprotein(a) bears little resemblance to the other apolipoproteins. One of its main subunits is a kringle, a structure that is widely found in the proteins of the fibrinolysis pathway.

The half-life of Lp(a) is 3.3 days, with cholesteryl ester as its main lipid. Similar to LDL, it contains apo B-100, which is linked with apo(a) by a disulfide bond. In vitro studies have shown that Lp(a) has a reduced affinity for the LDL receptor compared with LDL itself. Once taken up in the cell, Lp(a) may down-regulate de novo cholesterol synthesis. Lipoprotein(a) expression appears to be under genetic control, which determines the size of the Lp(a) macromolecule. The plasma concentration varies inversely with the molecular size of the Lp(a) subtype.

Epidemiologic evidence links elevated levels of Lp(a) with increased risk for CHD. The mechanisms for this risk are unclear. Modified Lp(a) may, like modified LDL, be taken up by the macrophage scavenger receptor pathway, thus contributing to the formation of foam cells, a key component of atherosclerotic plaque. Lipoprotein(a) displays homology with plasminogen, which is necessary to thrombolysis, and interference by Lp(a) with plasminogen activation may also help explain the enhanced risk associated with this lipoprotein. Lipoprotein(a) may prove to be an important intermediary in the relation between atherosclerosis and thrombosis.

High-Density Lipoprotein Metabolism

Several epidemiologic studies associate elevated HDL-C levels with reduced risk for CHD and, conversely, low levels of HDL-C with enhanced risk.[15-17] High-density lipoprotein mediates this benefit through several possible mechanisms, including reverse cholesterol transport, a process by which HDL removes cholesterol from peripheral cells and conveys it to the liver for biliary excretion. Reverse cholesterol transport does not involve a direct route from peripheral tissues to the liver, but depends on the repeated transfer of cholesteryl esters among lipoproteins (Figure 2-6). Other cardioprotective mechanisms of HDL may include inhibi-

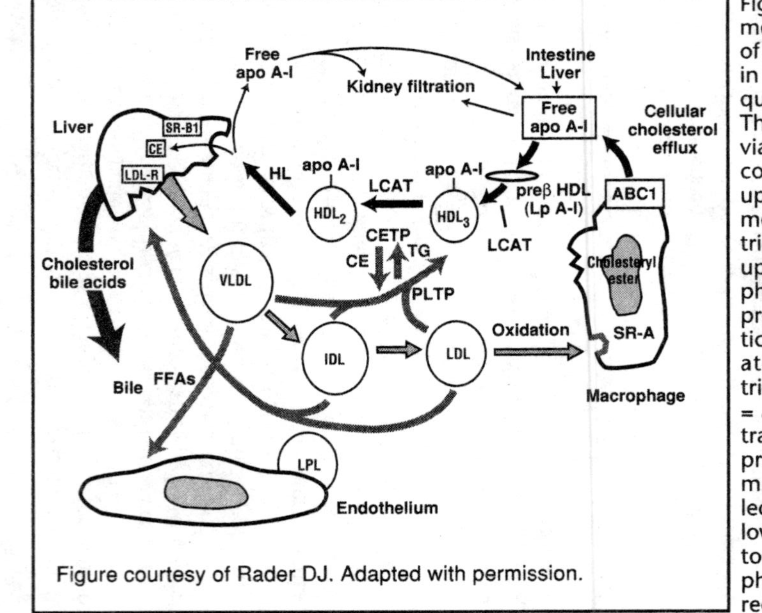

Figure courtesy of Rader DJ. Adapted with permission.

Figure 2-6: Pathway of HDL synthesis and metabolism. The synthesis and catabolism of HDL take place both in the intestine and in the liver. Nascent HDL (free apo A-I) acquires cholesterol from peripheral tissue. The cholesterol is then brought to the liver via two mechanisms: (1) LCAT-mediated conversion into HDL_3 and HDL_2, leading to uptake by the SR-B1 receptor; or (2) CETP-mediated exchange of cholesteryl ester for triglycerides from IDL and LDL, leading to uptake by the LDL receptor. PLTP transfers phospholipids from apo-B-containing lipoproteins to HDL. As part of this process, particles resembling preβ HDL (LpA-I) dissociate from the HDL. ABC1 = adenosine triphosphate (ATP) binding cassette 1; Apo = apolipoprotein; CETP = cholesteryl ester transfer protein; HDL = high-density lipoprotein; HL = hepatic lipase; IDL = intermediate-density lipoprotein; LCAT = lecithin:cholesterol acyltransferase; LDL = low-density lipoprotein; LDL-R = LDL receptor; LPL = lipoprotein lipase; PLTP = phospholipid transfer protein; SR-A = scavenger receptor A; SR-B1 = scavenger receptor B1.

tion of LDL oxidation[18] and effects on vasodilatation and platelet activation mediated by changes in the production of prostacyclin.[19] Interesting investigations have also explored the effect of inflammatory stimuli on the formation of a phenotype of HDL with reduced anti-atherogenicity, suggesting a link between inflammation and HDL functionality.[20-23]

Reverse Cholesterol Transport

The pathway of HDL synthesis and metabolism has become substantially better understood in recent years (Figure 2-6). High-density lipoprotein particles exhibit considerable diversity, as assessed by size, apolipoprotein and lipid constituents, and efficiency of clearance. One of the earliest precursors of HDL is a small apo A-I-rich particle that, as examined earlier, migrates to the preβ position on electrophoresis. In the early stages of reverse cholesterol transport, discoidal preβ-LpA-I accepts unesterified cholesterol from peripheral cells. The cholesterol-enriched particles are converted into small, spherical, α-migrating HDL_3, which is then remodeled into large, spherical particles (ie, HDL_2) whose lower mean weight ratio of apo A-I to cholesterol reflects their higher density.[6,24] These changes in composition and size, resulting in the formation of mature HDL, occur in part via the actions of lecithin:cholesterol acyltransferase (LCAT), which mediates the esterification of cholesterol, and phospholipid transfer protein (PLTP), which transfers phospholipids from apo-B-containing lipoproteins to HDL.[6] During this process, particles resembling preβ-LpA-I dissociate from the HDL.

Cholesterol is made available for biliary excretion in part through the actions of CETP and HL. CETP promotes the transfer of esterified cholesterol from large HDL particles to apo-B-containing lipoproteins in exchange for TG.[6] Triglyceride enrichment enhances the interaction of A-I HDL with PLTP. It also generates HDL particles that undergo HL-mediated hydrolysis and subsequent uptake by the scavenger receptor B1 in the liver. Some of the cholesteryl esters transferred to apo-B-containing lipoproteins are removed from the plasma via the LDL receptor in the liver. The hydrolysis of TG-enriched HDL particles is accompanied by the dissociation of lipid-poor apo A-I. Some of the lipid-poor particles are susceptible to renal excretion; others are formed into new discoidal preβ HDL.[6]

Role of Apolipoproteins

Evidence suggests that the protective effect of HDL-C is mediated in part by apo A-I.[25,26] In metabolic studies using radiolabeled lipoprotein particles, long-term exposure of different cells to LpA-I promoted cholesterol efflux; however, less efflux was observed during exposure to LpA-I:LpAI-AII.[27] The same researchers also found that increased expression of apo A-II in mice seems to promote aortic fatty streak development and decrease cholesterol efflux, while overexpression of apo A-I induced cholesterol efflux and protected against atherosclerosis.[27] Transgenic expression of human apo A-I in LDL-receptor-deficient mice is also associated with higher HDL-C levels compared with controls and with regression of established plaques by up to 70% compared with baseline.[28]

One study found that apo A-II-transgenic mice have traits associated with insulin resistance, including elevated levels of plasma FFA and TG, twofold elevations in plasma insulin levels compared with controls, and a delay in glucose clearance.[29] This is consistent with the hypothesis that insulin resistance and type 2 diabetes may sometimes be linked to alterations in lipid metabolism rather than glucose metabolism.[29]

Although these data support the hypothesis that apo A-II may be proatherogenic, other data suggest a more complex role. For example, human apo A-II transgenic mice fed an atherogenic diet have been found to develop less extensive aortic atherosclerosis than control mice, despite a dramatic decrease in endogenous apo A-I and A-II, together with a slight decrease in cholesterol efflux from peripheral cells.[30]

In another study, researchers investigated the binding of HDL to the scavenger receptor B1, which mediates the selective uptake of lipids from HDL into the liver and steroidogenic cells.[31-33] While the binding of reconstituted HDL (rHDL) containing apo A-I was twice greater than that of rHDL containing A-I/A-II, the selective uptake of cholesteryl ester from the bound particles was more efficient with the latter. These results suggest that apo A-II may have independent effects on ligand binding and cholesteryl ester uptake. An increase in the efficiency of uptake when apo A-II is present may indicate a positive effect on reverse cholesterol transport and a potentially antiatherogenic role of apo A-II.[31]

Structure and Function of HDL: Genetic Mutations

Several rare mutations affecting HDL structure or function have been identified in human populations, but data are inconsistent about their effects on cardiovascular risk. Tangier disease, first described 3 decades ago in a closed island community in the Chesapeake Bay, is characterized by the complete absence of normal HDL and by a low level of an abnormal HDL variant.[34] Nonetheless, early atherosclerosis is not a consistent feature of this condition.[35] Of recent interest has been the discovery of the molecular basis of Tangier disease in the original proband, who was homozygous for a deletion of nucleotides 3283 and 3284 in exon 22 of the ATP-binding-cassette A1 (ABCA1) gene.[36-38] Several different mutations in the ABCA1 gene have recently been described in French-Canadian and Dutch subjects, who also exhibited familial HDL deficiency. The ABCA1 gene encodes for cholesterol-efflux regulating protein (CERP).[39] A possible direction for future drug development may be the use of strategies such as retinoid X receptor agonists, which regulate ABCA1 expression, in order to change cholesterol balance.[40,41]

Other naturally occurring mutations that result in abnormal levels of HDL-C are similarly paradoxical. Subjects with fish-eye disease have markedly decreased levels, but no consistent increase in coronary risk.[42] Conversely, high levels of HDL are seen in individuals with CETP deficiency, but they may have increased coronary risk.[43]

In a rural area of Italy, individuals with low HDL-C levels and a variant of apo A-I have far less atherosclerosis than expected for their HDL-C concentrations. Characterized by the substitution of cysteine for arginine at position 173, allowing disulfide-linked dimer formation, the variant is known as apo A-I Milano. Recently, apo A-I Milano was synthesized in a phospholipid complex and administered by infusion to patients with acute coronary syndromes.[44] Compared with placebo, recombinant apo A-I Milano (ETC-216) in 5 weekly doses produced significant atherosclerosis regression as measured by intravascular ultrasound. The results of this small (N=123) trial need to be confirmed in larger trials that include clinical end points.

Turkish subjects appear to have genetically determined low HDL-C, both in Turkey and in Turkish expatriate populations in Europe. Despite this finding, in rural areas of Turkey there is little ischemic heart disease, presumably because of dietary and lifestyle factors. In urban areas, however, the low HDL-C is associated with an increased risk for coronary events, perhaps because of the presence of a more pro-atherogenic lifestyle.[45] Therefore, the atherogenicity of the low HDL-C state may be influenced by genetic and environmental factors.

Concepts in Atherogenesis

Theories regarding the antecedents of atherosclerosis are varied, and finding congruence among these viewpoints has proved to be controversial. The current consensus appears to support an inflammatory response-to-injury model that, in its broadest sense, embraces most perspectives on the origins of this disease.[46] According to this model, atherogenesis is a response to injury of the vascular endothelium. At its mildest, atherogenesis may be a reparative process that leads to clinically insignificant thrombus formation and smooth muscle cell proliferation that covers the site of endothelial damage. At its most severe, atherogenesis may lead to the development of lesions that contribute to clinical syndromes, eg, unstable angina pectoris or myocardial infarction. Vascular endothelium may be damaged by cytotoxic molecules or by alterations in blood flow resulting from changes in arterial topography. This initial injury may not necessarily result in endothelial denudation, but rather a derailment of normal endothelial function. Histologic studies have determined that early lesions develop at sites of morphologically intact endothelium.

Dysfunctional Endothelium, Monocyte Adhesion, and Smooth Muscle Cell Activity

Atherogenesis may be described as an immunologic, inflammatory response to endothelial insult.[47] Damaged or dysfunctional endothelium is prothrombotic and chemotactic to monocytes that migrate to the subendothelial, or intimal, space under the stimulation of various cytokines and adhesion molecules. Monocytes may infiltrate the intima at the site of injury or through junctions between endothelial cells.[48] Within the intima, these cells differentiate into macrophages that may promote both lipid oxidation and uptake of oxidatively modified lipoproteins by the scavenger receptor. Macrophages become the lipid-laden foam cells that characterize the fatty streak, the earliest atherosclerotic lesion. Foam cell death may account for the intimal deposition of a necrotic pool of lipid, characteristic of certain atherosclerotic lesions (Figure 2-7).

Smooth muscle cells may proliferate over the early atherosclerotic lesion and develop into the fibrous cap that separates the contents of the lesion from the plasma compartment.[49] Smooth muscle cells may also take up oxidized LDL in the intima. Progression of atherosclerotic disease may be accounted for by a mechanism of repeated plaque rupture, superficial thrombosis, and smooth muscle cell proliferation.

Macrophages play a second role in atherogenesis through their interaction with smooth muscle cells. Macrophage activity appears most heavily localized around the shoulders of the fibrous cap. Through a variety of cytokines, proteinases, and growth factors, macrophage activity can influence smooth muscle cell proliferation, as well as smooth muscle cell production of collagen and other matrix proteins that preserve the struc-

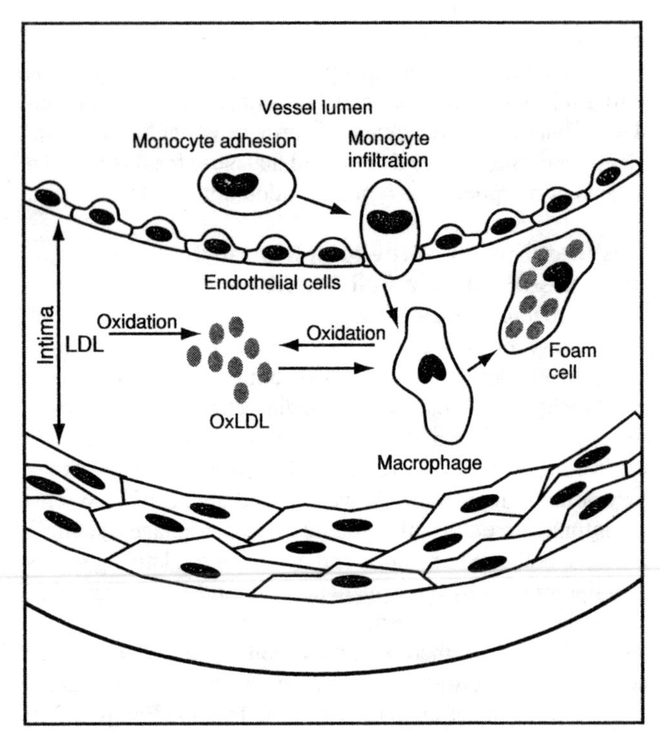

Figure 2-7: Foam cell formation. Foam cells characterize the earliest atherosclerotic lesion, the fatty streak. OxLDL= oxidized low-density lipoprotein.

tural integrity of the fibrous cap. These effects are particularly important in considering stabilization of rupture-prone lesions.

Rupture of Atherosclerotic Lesions

Most coronary ischemic events are thrombotic, originating from the rupture of the atherosclerotic lesion.[50] Identifying those plaques susceptible to rupture has important therapeutic implications. Generally, mature plaques consist of soft, lipid-rich atheromatous 'gruel' and sclerotic tissue that is hard, fibrous, and rich in collagen. Although the sclerotic component constitutes >70% of the average stenotic coronary plaque, it appears to stabilize plaques and protect them from rupture. The soft, lipid-rich component is more dangerous because it destabilizes plaques and makes them vulnerable to rupture. Once rupture occurs, the thrombogenic constituents of the plaque core are exposed to flowing blood, which can result in thrombosis.[51]

In patients with ischemia caused by vascular obstruction, tissue perfusion may continue because of the development of collateral vessels.[52] An evaluation of 106 patients, 21% of whom (n = 22) presented with an acute coronary syndrome and 43% of whom (n = 46) had experienced a prior acute coronary syndrome, indicated that collateral vessel development was more likely in patients with stenoses >90%.[53] This characteristic could serve as the basis for the development of angiogenic therapy to enhance collateral flow.[52] Autopsy studies have characterized the morphologic features of vulnerable lesions: a weakened fibrous cap that may be an inflamed, focal accumulation of macrophage foam cells, a lipid-rich core, and few smooth muscle cells.[54,55] The activated macrophage can secrete several matrix metalloproteinases that can lead to the hydrolysis of collagen and fibrin and disrupt the fibrous cap.

Therapeutic Mechanisms

Clinical trials showing the beneficial effects of aggressive lipid regulation on progression of atherosclerotic disease have reported that modest changes in vascular end points correspond to disproportionately large reductions in clinical events. Several mechanisms have been postulated by which lipid-lowering therapy may mediate this benefit.[56] These 'pleiotropic' mechanisms include restoration of endothelial function, improvement in myocardial ischemia, reduction in macrophage and inflammatory activity, antioxidant effects that prevent lipoprotein modification, and reduction in platelet thrombogenicity. Some of the pleiotropic effects of statin therapy may be direct and independent of lipid lowering.

Restoration of Endothelial Function

The endothelial cells that line the vessel wall produce a variety of molecules that maintain vasomotor tone through control of smooth muscle cell contraction and relaxation. A key molecule is nitric oxide, a powerful endothelium-derived vasodilator, or one of its analogs. However, damage to the endothelial cells by oxidized lipoproteins or a host of other cytotoxic agents, and the presence of CHD, may disturb the normal function of these molecules. Endothelium-independent vasodilators (eg, nitroglycerin) can bypass the endothelium, exerting their influence directly on smooth muscle cells.

Endothelium-dependent vasodilators, such as nitric oxide and acetylcholine, induce a paradoxical vasoconstrictive response in atherosclerotic vessels. This endothelial dysfunction may hamper the vessel's ability to respond to stress and may be important in precipitating the clinical manifestation of CHD. Several trials, including the Reduction of Cholesterol in Ischemia and Function of the Endothelium (RECIFE) study, have shown that lipid regulation can improve vasodilatation.[57-61] The RECIFE study enrolled patients hospitalized for an acute myocardial infarction (MI) or unstable angina.[61] Results of this trial indicate that pravastatin (Pravachol®) 40 mg/d vs placebo produced a 42% increase from baseline in endothelium-dependent vasodilatation (P = 0.02).[61] There was no significant correlation between the percent improvement in flow-mediated dilatation and the decrease in TC or LDL-C. Clinical trials are needed to determine whether improved vasodilatation is associated with reductions in cardiovascular events.

Studies of lipid regulation and vasodilatation raise a number of important questions, including whether improvement in blood flow is a class effect of statins and whether it is more likely to occur in certain groups of patients. For example, in one randomized double-blind study, subjects with coronary artery disease who received simvastatin (Zocor®) 40 mg/d for 6 months had no significant improvement in vasomotor function.[62]

Reduced Ischemia

An improvement in endothelial function may translate into improved blood-flow response to myocardial demand. In a study by Gould et al, an aggressive lipid-regulating program of diet and lifestyle modification yielded significantly enhanced myocardial perfusion, as assessed by positron emission tomography scanning.[63] Investigations have also reported fewer incidents of ischemic episodes with lipid-lowering therapy, as measured by ambulatory Holter monitoring. In the Myocardial Ischemia Reduction with Aggressive Cholesterol Lowering (MIRACL) trial, post-MI patients receiving atorvastatin (Lipitor®) 80 mg/d experienced a 16% reduction in primary end point events (death, nonfatal acute MI, cardiac arrest, recurrent symptomatic myocardial ischemia requiring rehospitalization) ($P = 0.048$) that was primarily the result of a 26% reduction in myocardial ischemia ($P = 0.02$).[64]

Reduced Inflammatory and Immune Activity

Collagen content is critical to the tensile strength and stability of atherosclerotic plaque, and collagen accumulation in complex atheroma seems to correlate with the number of smooth muscle cells.[65,66] Plaque instability may result in part from the infiltration of coronary lesions by immune cells, including macrophages and T cells.[67] When activated, macrophages secrete proteolytic enzymes (eg, matrix metalloproteinases) that can weaken the collagen-containing fibrous cap separating the lipid core of the lesion from the bloodstream. Weakening of the fibrous cap leads to plaque rupture and thrombosis.[66] Migration of certain T cells into plaques can precipitate tissue-destructive events, including IFN-γ-mediated activation of metalloproteinase-secreting macrophages, smooth muscle cell cytolysis, and endothelial cell damage.[67]

Research suggests that lipid lowering can contribute to plaque stability by decreasing macrophage accumulation in atherosclerotic lesions and inhibiting production of matrix metalloproteinases by activated macrophages.[66] In a rabbit model, it has also been shown that cholesterol lowering with pravastatin, a hydrophilic statin, is associated with more smooth muscle cells and collagen accumulation compared with placebo.[65] In addition, experimental evidence suggests that statins may inhibit T-cell proliferation and suppress the secretion of IFN-γ from T cells.[68] Some effects of lipid-lowering statins may be cholesterol independent and occur earlier than cholesterol-dependent effects.[66]

On a molecular level, researchers are investigating the role of peroxisome proliferator-activated receptor γ (PPARγ), a nuclear transcription factor expressed in macrophages and other major cell types involved in vascular injury. Activation of PPARγ by endogenous or synthetic ligands may inhibit matrix metalloproteinase activity and have other antiatherogenic effects.[69,70]

Antioxidant Effect of Lipid Regulation

Because oxidized lipoproteins are thought to be pivotal in the initial endothelial insult that precedes lesion formation, inhibiting the production of oxidized particles may be desirable. In an analysis of lipoprotein particles obtained from patients in the Kuopio Atherosclerosis Prevention Study (KAPS), treatment with pravastatin alone was found to have reduced the oxidative capacity of LDL.[71] Although the recent Heart Protection Study (N=20,536) found that the antioxidant vitamins E, C, and β-carotene, administered with and without simvastatin, did not reduce cardiovascular risk in a range of high-risk subjects,[72] there are multiple oxidative pathways. Some experts have postulated that statins exert an antioxidative effect via pathways relevant to the atherogenic process.

Reduced Platelet Thrombogenicity

Blood from patients with hypercholesterolemia has increased thrombotic potential compared with that from normocholesterolemic controls. In two studies, thrombus formation was assessed ex vivo in hypercholesterolemic patients with coronary disease who were treated with pravastatin.[73,74] Compared with baseline, platelet thrombus formation decreased, an effect that may be linked to reductions in cholesterol levels. One of the studies also enrolled subjects without coronary disease, and a greater decrease from baseline in thrombus formation was found in this group. However, the increase in fibrinogen levels that occurred in the latter study illustrates the complexity of the effect that statins may have on thrombosis.[74]

References

1. Baggott J, Dennis SE: Macromolecules. ©1994, 1995. Available at: http://medlib.med.utah.edu/NetBiochem/macromol.htm. Accessed November 1, 2004.

2. Denniston KJ: *General, Organic, and Biochemistry*, 3rd ed. New York, NY, McGraw-Hill, 2000, p 516.

3. Gotto AM Jr, Pownall H: *Manual of Lipid Disorders*, 3rd ed. Philadelphia, Lippincott Williams & Wilkins, 2003, pp 2-65.

4. Castellani LW, Lusis AJ: ApoA-II versus apoA-I: two for one is not always a good deal [editorial]. *Arterioscler Thromb Vasc Biol* 2001;21:1870-1872.

5. Lee J, von Eckardstein A, Lindstedt L, et al: Depletion of pre-β LpA1 and LpA4 particles by mast cell chymase reduces cholesterol efflux from macrophage foam cells induced by plasma. *Arterioscler Thromb Vasc Biol* 1999;19:1066-1074.

6. Rye KA, Clay MA, Barter PJ: Remodelling of high density lipoproteins by plasma factors. *Atherosclerosis* 1999;145:227-238.

7. von Eckardstein A, Nofer JR, Assmann G: Acceleration of reverse cholesterol transport. *Curr Opin Cardiol* 2000;15:348-354.

8. Genest JJ Jr, Bard JM, Fruchart JC, et al: Plasma apolipoprotein A-I, A-II, B, E and C-III containing particles in men with premature coronary artery disease. *Atherosclerosis* 1991;90:149-157.

9. Raiha I, Marniemi J, Puukka P, et al: Effect of serum lipids, lipoproteins, and apolipoproteins on vascular and nonvascular mortality in the elderly. *Arterioscler Thromb Vasc Biol* 1997;17:1224-1232.

10. Gotto AM Jr, Whitney E, Stein EA, et al: Relation between baseline and on-treatment lipid parameters and first acute major coronary events in the Air Force/Texas Coronary Atherosclerosis Prevention Study (AFCAPS/TexCAPS). *Circulation* 2000;101:477-484.

11. Berenson GS, Srinivasan SR, Wattigney W, et al: Insight into a bad omen for white men: coronary artery disease—the Bogalusa Heart Study. *Am J Cardiol* 1989;64:32C-39C.

12. Brown MS, Goldstein JL: Familial hypercholesterolemia: defective binding of lipoproteins to cultured fibroblasts associated with impaired regulation of 3-hydroxy-3-methylglutaryl coenzyme A reductase activity. *Proc Natl Acad Sci USA* 1974;71:788-792.

13. Cohn JS, Marcoux C, Davignon J: Detection, quantification, and characterization of potentially atherogenic triglyceride-rich remnant lipoproteins. *Arterioscler Thromb Vasc Biol* 1999;19:2474-2486.

14. Loscalzo J: Lipoprotein(a). A unique risk factor for atherothrombotic disease. *Arteriosclerosis* 1990;10:672-679.

15. Castelli WP, Doyle TG, Hames CG, et al: HDL and other lipids in coronary heart disease: the cooperative lipoprotein phenotyping study. *Circulation* 1977;55:767-772.

16. Barter PJ, Rye KA: High density lipoproteins and coronary heart disease. *Atherosclerosis* 1996;121:1-12.

17. Gordon DJ: HDL and coronary heart disease—an epidemiological perspective. *J Drug Dev* 1993;3(suppl 1):11-17.

18. Graham A, Hassal DG, Rafique S, et al: Evidence for paraoxonase-independent inhibition of low-density lipoprotein oxidation by high-density lipoprotein. *Atherosclerosis* 1997;135:193-204.

19. Myers DE, Huang WN, Larkins RG: Lipoprotein-induced prostacyclin production in endothelial cells and effects of lipoprotein modification. *Am J Physiol* 1996;271:C1504-C1511.

20. Khovidhunkit W, Shigenaga JK, Moser AH, et al: Cholesterol efflux by acute-phase high density lipoprotein: role of lecithin: cholesterol acyltransferase. *Lipid Res* 2001;42:967-975.

21. Navab M, Berliner JA, Subbanagounder G, et al: HDL and the inflammatory response induced by LDL-derived oxidized phospholipids. *Arterioscler Thromb Vasc Biol* 2001;21:481-488.

22. Nofer JR, Kehrel B, Fobker M, et al: HDL and arteriosclerosis: beyond reverse cholesterol transport. *Atherosclerosis* 2002;161:1-16.

23. Libby P: Managing the risk of atherosclerosis: the role of high-density lipoprotein. *Am J Cardiol* 2001;88(suppl):3N-8N.

24. Patsch W, Brown SA, Morrisett JD, et al: A dual-precipitation method evaluated for measurement of cholesterol in high-density lipoprotein subfractions HDL_2 and HDL_3 in human plasma. *Clin Chem* 1989;35:265-270.

25. Plump AS, Scott CJ, Breslow JL: Human apolipoprotein A-1 gene expression increases high density lipoprotein and suppresses

atherosclerosis in the apolipoprotein E-deficient mouse. *Proc Natl Acad Sci USA* 1994;91:9607-9611.

26. Paszty C, Maeda N, Verstuyft J, et al: Apolipoprotein A1 transgene corrects apolipoprotein E deficiency-induced atherosclerosis in mice. *J Clin Invest* 1994;94:899-903.

27. Fruchart JC, De Geteire C, Delfly B, et al: Apolipoprotein A-I-containing particles and reverse cholesterol transport: evidence for connection between cholesterol efflux and atherosclerosis risk. *Atherosclerosis* 1994;110(suppl):S35-S39.

28. Tangirala RK, Tsukamoto K, Chun SH, et al: Regression of atherosclerosis induced by liver-directed gene transfer of apolipoprotein A-1 in mice. *Circulation* 1999;100:1816-1822.

29. Castellani LW, Gotto AM, Lusis AJ: Studies with apolipoprotein A-II transgenic mice indicate a role for HDLs in adiposity and insulin resistance. *Diabetes* 2001;50:643-651.

30. Tailleux A, Bouly M, Luc G, et al: Decreased susceptibility to diet-induced atherosclerosis in human apolipoprotein A-II transgenic mice. *Arterioscler Thromb Vasc Biol* 2000;20:2453-2458.

31. de Beer MC, Durbin DM, Cai L, et al: Apolipoprotein A-II modulates the binding and selective lipid uptake of reconstituted high-density lipoprotein by scavenger receptor BI. *J Biol Chem* 2001;276:15832-15839.

32. Trigatti B, Rigotti A, Krieger M: The role of the high-density lipoprotein receptor SR-BI in cholesterol metabolism. *Curr Opin Lipidol* 2000;11:123-131.

33. Trigatti B, Rayburn H, Vinals M, et al: Influence of the high-density lipoprotein receptor SR-BI on reproductive and cardiovascular pathophysiology. *Proc Natl Acad Sci USA* 1999;96:9322-9827.

34. Lux SE, Levy RI, Gotto AM, et al: Studies on the protein defect in Tangier disease. Isolation and characterization of an abnormal high density lipoprotein. *J Clin Invest* 1972;51:2505-2519.

35. Serfaty-Lacrosniere C, Civeira F, Lanzberg A, et al: Homozygous Tangier disease and cardiovascular disease. *Atherosclerosis* 1994;107:85-98.

36. Brooks-Wilson A, Marcil M, Clee SM, et al: Mutations in ABC1. *Nat Genet* 1999;22:336-345.

37. Rust S, Rosier M, Funke H, et al: Tangier disease is caused by mutations in the gene encoding ATP-binding cassette transporter 1. *Nat Genet* 1999;22:352-355

38. Bodzioch M, Orso E, Klucken J, et al: The gene encoding ATP-binding cassette transporter 1 is mutated in Tangier disease. *Nat Genet* 1999;22:347-351.

39. Marcil M, Brooks-Wilson A, Clee SM, et al: Mutations in the ABC1 gene in familial HDL deficiency with defective cholesterol efflux. *Lancet* 1999;354:1341-1346.

40. Ferber D: Lipid research. Possible new way to lower cholesterol. *Science* 2000;289:1446-1447.

41. Repa JJ, Turley SD, Lobaccaro JA, et al: Regulation of absorption and ABC1-mediated efflux of cholesterol by RXR heterodimers. *Science* 2000;289:1524-1529.

42. Kuivenhoven JA, van Voorst EJ, Wiebusch H, et al: A unique genetic and biochemical presentation of fish-eye disease. *J Clin Invest* 1995;96:2783-2791.

43. Zhong S, Sharp DS, Grove JS, et al: Increased coronary artery disease in Japanese-American men with mutation in the cholesteryl ester transfer protein gene despite increased HDL levels. *J Clin Invest* 1996;97:2917-2923.

44. Nissen SE, Tsunoda T, Tuzcu EM, et al: Effect of recombinant ApoA-I Milano on coronary atherosclerosis in patients with acute coronary syndromes: a randomized controlled trial. *JAMA* 2003; 290:2292-2300.

45. Mahley RW, Palaoglu KE, Atak Z, et al: Turkish Heart Study: lipids, lipoproteins and apolipoproteins. *J Lipid Res* 1995;36:839-859.

46. Ross R: The pathogenesis of atherosclerosis: a perspective for the 1990s. *Nature* 1993;362:801-809.

47. Libby P: Current concepts of the pathogenesis of the acute coronary syndromes. *Circulation* 2001;104:365-372.

48. Davies MJ: Pathogenesis of atherosclerosis. *Curr Opin Cardiol* 1992;7:541-545.

49. Weissberg PL, Clesham GJ, Bennett MR: Is vascular smooth muscle cell proliferation beneficial? *Lancet* 1996;347:305-307.

50. Brown BG: Lipid-lowering therapy for the stabilization of the vulnerable atherosclerotic plaque. *Curr Opin Lipidol* 1993;4:305-309.

51. Falk E, Shah PK, Fuster V: Coronary plaque disruption. *Circulation* 1995;92:651-671.

52. Epstein SE, Fuchs S, Zhou YF, et al: Therapeutic interventions for enhancing collateral development by administration of

growth factors: basic principles, early results and potential hazards. *Cardiovasc Res* 2001;49:532-542.

53. Mason MJ, Walker SK, Patel DJ, et al: Influence of clinical and angiographic factors on development of collateral channels. *Coron Artery Dis* 2000;11:573-578.

54. Conti CR: Updated pathophysiologic concepts in unstable coronary artery disease. *Am Heart J* 2001;141(2 suppl):S12-S14.

55. Richardson PD, Davies MJ, Born GV: Influence of plaque configuration and stress distribution on fissuring of coronary atherosclerotic plaques. *Lancet* 1989;2:941-944.

56. Levine GN, Keaney JF Jr, Vita JA: Cholesterol reduction in cardiovascular disease. Clinical benefits and possible mechanisms. *N Engl J Med* 1995;332:512-521.

57. Egashira K, Hirooka Y, Kai H, et al: Reduction in serum cholesterol with pravastatin improves endothelium-dependent coronary vasomotion in patients with hypercholesterolemia. *Circulation* 1994;89:2519-2524.

58. Treasure CB, Klein JL, Weintraub WS, et al: Beneficial effects of cholesterol-lowering therapy on the coronary endothelium in patients with coronary artery disease. *N Engl J Med* 1995; 332:481-487.

59. Anderson TJ, Meredith IT, Yeung AC, et al: The effect of cholesterol-lowering and antioxidant therapy on endothelium-dependent coronary vasomotion. *N Engl J Med* 1995;332:488-493.

60. John S, Schlaich M, Langenfeld M, et al: Increased bioavailability of nitric oxide after lipid-lowering therapy in hypercholesterolemic patients: a randomized, placebo-controlled, double-blind study. *Circulation* 1998;98:211-216.

61. Dupuis J, Tardif JC, Cernacek P, et al: Cholesterol reduction rapidly improves endothelial function after acute coronary syndromes. The RECIFE (Reduction of Cholesterol in Ischemia and Function of the Endothelium) Trial. *Circulation* 1999;99:3227-3233.

62. Vita JA, Yeung AC, Winniford M, et al: Effect of cholesterol-lowering therapy on coronary endothelial vasomotor function in patients with coronary artery disease. *Circulation* 2000;102:846-851.

63. Gould KL, Ornish D, Scherwitz L, et al: Changes in myocardial perfusion abnormalities by positron emission tomography after long-term, intense risk factor modification. *JAMA* 1995;274:894-901.

64. Schwartz GG, Olsson AG, Ezekowitz MD, et al: Effects of atorvastatin on early recurrent ischemic events in acute coronary

syndromes: the MIRACL study: a randomized controlled trial. *JAMA* 2001;285:1711-1718.

65. Fukumoto Y, Libby P, Rabkin E, et al: Statins alter smooth muscle cell accumulation and collagen content in established atheroma of Watanabe heritable hyperlipidemic rabbits. *Circulation* 2001;103:993-999.

66. Takemoto M, Liao JK: Pleiotropic effects of 3-hydroxy-3-methylglutaryl coenzyme A reductase inhibitors. *Arterioscler Thromb Vasc Biol* 2001;21:1712-1719.

67. Liuzzo G, Kopecky SL, Frye RL, et al: Perturbation of the T-cell repertoire in patients with unstable angina. *Circulation* 1999;100:2135-2139.

68. Tanaka T, Soejima H, Hirai N, et al: Comparison of frequency of interferon-γ-positive CD4+ T cells before and after percutaneous coronary intervention and the effect of statin therapy in patients with stable angina pectoris. *Am J Cardiol* 2004;93:1547-1549.

69. Collins AR, Meehan AP, Kintscher U, et al. Troglitazone inhibits formation of early atherosclerotic lesions in diabetic and nondiabetic low density lipoprotein receptor-deficient mice. *Arterioscler Thromb Vasc Biol* 2001;21:365-371.

70. Marx N, Sukhova G, Murphy C, et al. Macrophages in human atheroma contain PPARγ: differentiation-dependent peroxisomal proliferator-activated receptor γ (PPARγ) expression and reduction of MMP-9 activity through PPARγ activation in mononuclear phagocytes in vitro. *Am J Pathol* 1998;153:17-23.

71. Salonen R, Nyssonen K, Porkkala-Sarataho E, et al: The Kuopio Atherosclerosis Prevention Study (KAPS): effect of pravastatin treatment on lipids, oxidation resistance of lipoproteins, and atherosclerotic progression. *Am J Cardiol* 1995;76:34C-39C.

72. MRC/BHF Heart Protection Study of antioxidant vitamin supplementation in 20,536 high-risk individuals: a randomised placebo-controlled trial. *Lancet* 2002;360:23-33.

73. Lacoste L, Lam JY, Hung J, et al: Hyperlipidemia and coronary disease. Correction of the increased thrombogenic potential with cholesterol reduction. *Circulation* 1995;92:3172-3177.

74. Dangas G, Badimon JJ, Smith DA, et al: Pravastatin therapy in hyperlipidemia: effects on thrombus formation and the systemic hemostatic profile. *J Am Coll Cardiol* 1999;33:1294-1304.

Chapter 3

Risk Assessment and Risk Factor Reduction

Two approaches are generally considered for the prevention of coronary heart disease (CHD). The first is a population-wide effort to provide education about lifestyle modification. The second is to identify and evaluate CHD risk in individual patients in the clinical setting.

National educational initiatives have made headway in the first objective, while numerous groups have achieved progress in the second objective by issuing guidelines to manage CHD risk in individual patients.

Because this handbook focuses on the management of lipid disorders, the emphasis has been placed on the guidelines established by the Adult Treatment Panel III (ATP III) of the National Cholesterol Education Program (NCEP), including risk stratification and interventions based on a patient's lipid profile and associated risk factors.[1,2] However, increasing attention to the synergistic effect of multiple risk factors has highlighted the need to broaden our understanding of risk (see Concept of Global Risk, this chapter).

The guidelines issued by ATP III affirm that the primary goal for decreasing CHD risk is to reduce low-density lipoprotein cholesterol (LDL-C) levels.[1,2] This principle was also fundamental to earlier NCEP guidelines.[3]

Patients with established CHD, other forms of atherosclerotic disease, and diabetes are placed in a high-risk category that requires a minimum LDL-C goal of <100 mg/dL. For very high-risk patients, an optional LDL-C goal of <70 mg/dL can be considered. Very high risk is defined as the presence of established cardiovascular disease (CVD) plus multiple major risk factors, severe or poorly controlled risk factors, multiple risk factors of the metabolic syndrome, or an acute coronary syndrome.

When none of the above conditions are present, the risk category and LDL-C goal can be determined by calculating the patient's absolute risk for CHD using a scoring system developed from the Framingham Heart Study. Begun in 1948 by the National Heart Institute (now the National Heart, Lung, and Blood Institute [NHLBI]), the Framingham Heart Study was designed to identify risk factors for CVD by following disease development over an extended period in a large number of subjects who were asymptomatic at the start of the study.[4]

To calculate absolute risk for CHD, the physician must first conduct a comprehensive risk assessment that takes into account both lipid and nonlipid risk factors (Figure 3-1). This comprehensive risk assessment must include a fasting lipoprotein profile, which should be obtained every 5 years for all adults 20 years of age or older. The fasting lipoprotein profile reports:

- total cholesterol (TC);
- LDL-C;
- high-density lipoprotein cholesterol (HDL-C);
- triglycerides (TGs).

The LDL-C level can be estimated by the Friedewald equation: LDL-C = TC – HDL-C – (TG/5). Use of the Friedewald equation requires a fasting measurement of three lipid subfractions: TC, HDL-C, and TG levels. Furthermore, TG levels must be <400 mg/dL to prevent miscalculation caused by very-low-density lipoprotein (VLDL) particles of abnormal composition. When TG is ≥400 mg/dL, it is more accurate and far more desirable to measure the LDL-C concentration using direct ultracentrifugation.

Despite the importance of the fasting lipoprotein profile, it is just one of several elements that are essential to a comprehensive risk assessment. Each of the crucial elements is reviewed below.

Comprehensive Risk Assessment

The Medical Interview and Physical Examination

Risk assessment begins with the medical interview, which is the foundation of medical care and the clinician's most important activity.[5] The purpose of the medical interview is twofold: (1) to obtain critical information, and (2) to establish effective physician-patient communication.

The algorithm in Figure 3-1 presents the information that the physician must obtain during the medical interview. This information will help identify the presence of the following major CHD risk factors specified by ATP III:

- cigarette smoking;
- hypertension (blood pressure ≥140/90 mm Hg or use of antihypertensive medication);
- low HDL-C (<40 mg/dL);
- family history of premature CHD in first-degree male (<55 years of age) or female (<65 years of age) relative;
- age (male ≥45 years of age, female ≥55 years of age).

The interview also gives the physician an opportunity to learn about patient beliefs and behaviors that can influence compliance with treatment.[6] Because positive physician-patient interaction is associated with adherence to lipid-lowering therapy,[7] busy physicians are urged to spend the time needed to obtain patient information and to establish the rapport that can have a positive effect on the entire therapeutic relationship.

A complete cardiovascular examination at the time of the initial medical interview is also essential. This should include a measurement of waist circumference in obese patients. Abdominal obesity has been implicated in the development of insulin resistance, a metabolic disorder linked to a constellation of lipid and nonlipid risk factors known as the metabolic syndrome. Collectively, the characteristics of the metabolic syndrome increase the risk for CHD at any LDL-C level. There-

Figure 3-1: Algorithm Based on the Third Report of the National Cholesterol Education Program (NCEP) Adult Treatment Panel III (ATP III) and the 2004 Report on the Implications of Recent Clinical Trials

1. Patient History and Complete Cardiovascular Examination

Age*/ Gender	**Height/Weight** **Waist circumference** • Abdominal obesity - >88 cm (>35 in) female - >102 cm (>40 in) male	**Lifestyle** • Daily diet - saturated fat - cholesterol - carbohydrates - calories • Tobacco use* • Physical activity - 30-60 min/d or ≥3 x/wk - aerobic exercise (eg, walking, cycling, jogging) • Nature of occupation (eg, sedentary)

* CHD risk factor

1. Patient History and Complete Cardiovascular Examination *(continued)*

Medical				
Lipids • High LDL-C • Low HDL-C* • High TG	**Hypertension***	**CHD/Atherosclerosis** **Patient; Family (premature*)** • MI • Angina • CABG • PCI • PAD** • AAA** • Carotid artery disease** **Family (premature) CHD*** • <55 yr in first-degree male relative • <65 yr in first-degree female relative	**Diabetes**/ Hyperglycemia**	**Medication Use** • Antihypertensive medications* • Medications causing secondary dyslipidemia • Other

* CHD risk factor
** CHD risk equivalent

LDL-C = low-density lipoprotein cholesterol
HDL-C = high-density lipoprotein cholesterol

TG = triglyceride
CHD = coronary heart disease
MI = myocardial infarction
CABG = coronary artery bypass graft
PCI = percutaneous coronary intervention
PAD = peripheral arterial disease
AAA = abdominal aortic aneurysm

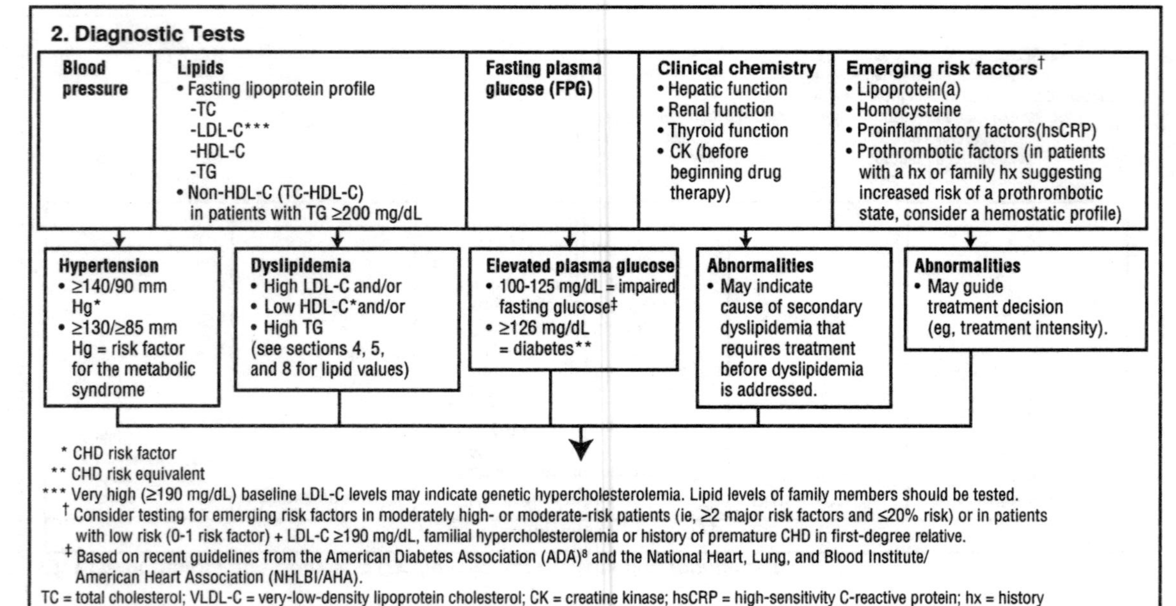

* CHD risk factor

** CHD risk equivalent

*** Very high (≥190 mg/dL) baseline LDL-C levels may indicate genetic hypercholesterolemia. Lipid levels of family members should be tested.

† Consider testing for emerging risk factors in moderately high- or moderate-risk patients (ie, ≥2 major risk factors and ≤20% risk) or in patients with low risk (0-1 risk factor) + LDL-C ≥190 mg/dL, familial hypercholesterolemia or history of premature CHD in first-degree relative.

‡ Based on recent guidelines from the American Diabetes Association (ADA)[8] and the National Heart, Lung, and Blood Institute/American Heart Association (NHLBI/AHA).

TC = total cholesterol; VLDL-C = very-low-density lipoprotein cholesterol; CK = creatine kinase; hsCRP = high-sensitivity C-reactive protein; hx = history

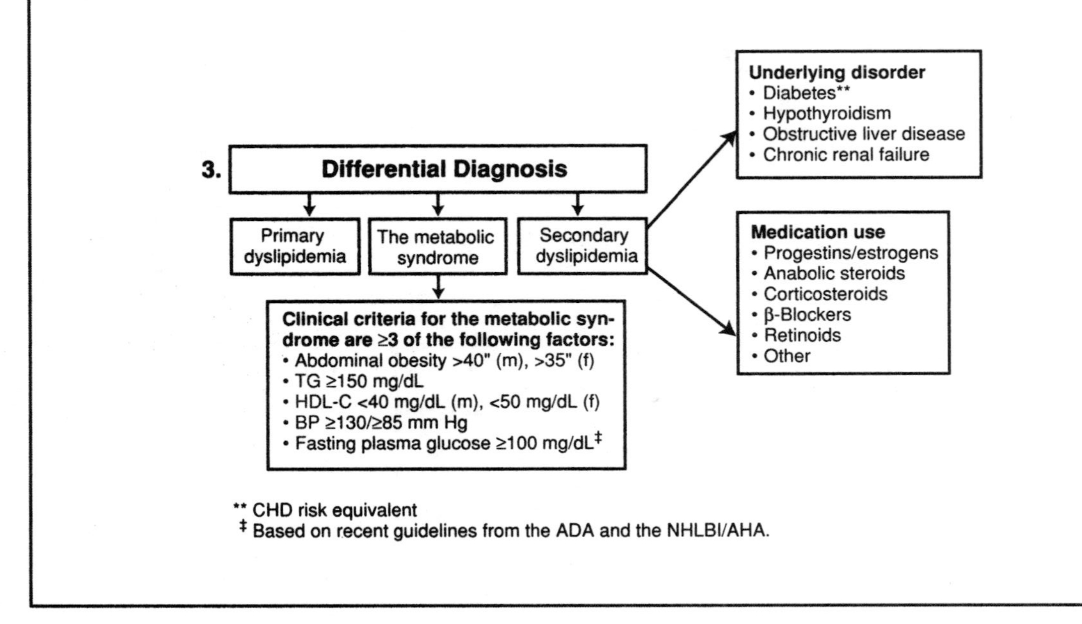

** CHD risk equivalent
‡ Based on recent guidelines from the ADA and the NHLBI/AHA.

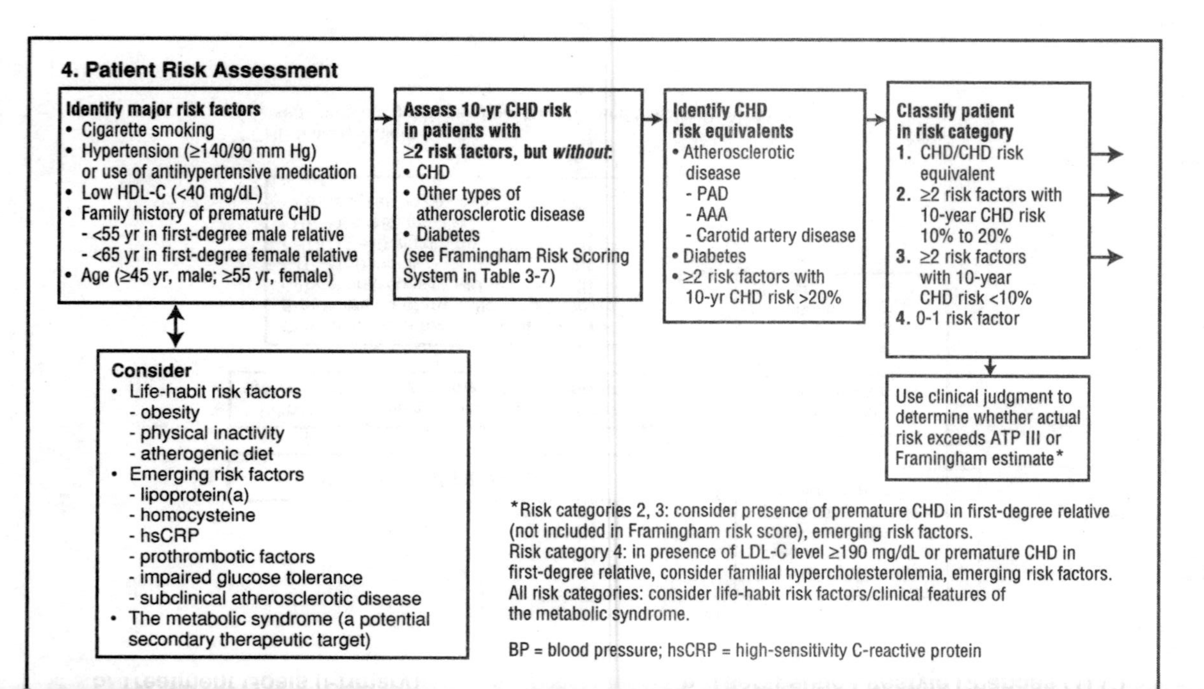

*Risk categories 2, 3: consider presence of premature CHD in first-degree relative (not included in Framingham risk score), emerging risk factors.
Risk category 4: in presence of LDL-C level ≥190 mg/dL or premature CHD in first-degree relative, consider familial hypercholesterolemia, emerging risk factors.
All risk categories: consider life-habit risk factors/clinical features of the metabolic syndrome.

BP = blood pressure; hsCRP = high-sensitivity C-reactive protein

5. Treatment Goals (Primary)

Identify LDL-C cut points/goals (mg/dL) for therapy

Risk Category	Cut point: *TLC*	Cut point: *Drug Therapy*	Goal
1	≥100**	≥100***† (<100: consider drug options- very high risk)††	<100 (<70: optional goal- very high risk)††
2 10% to 20% risk	≥130**	≥130*** (100-129: consider drug options)	<130 (<100: optional goal)
3 <10% risk	≥130	≥160	<130
4	≥160	≥190‡ (160-189: consider drug options)	<160

6. Therapeutic Lifestyle Changes (TLC)

Begin TLC§

6 weeks

Monitor LDL-C response/emphasize adherence
- Intensify TLC as required
  - add plant stanols/sterols
  - increase fiber intake
  - refer for medical nutrition therapy

6 weeks

Monitor response/emphasize adherence
As needed:
- Consider drug options§
- Treat the metabolic syndrome
  - weight loss
  - physical activity
- Refer for medical nutrition therapy

to 7

4 to 6 months

Monitor response/emphasize adherence

** Initiate TLC regardless of LDL-C level in patients with lifestyle risk factors.
*** Based on clinical judgment, consider doses sufficient to lower LDL-C by 30% to 40% (within the bounds of safety/tolerability) to achieve at least the minimum target (in Category 1 patients with very high baseline LDL-C, reduction to <100 mg/dL may not be possible).
† In high-risk patients with high TG or low HDL-C, consider combining a fibrate or nicotinic acid with an LDL-lowering drug.
†† Very high risk: established CVD + (1) multiple major risk factors (eg, diabetes), (2) severe/poorly controlled risk factors (eg, smoking), (3) metabolic syndrome, (4) acute coronary syndromes.
‡ Very high (≥190 mg/dL) baseline LDL-C levels may indicate genetic hypercholesterolemia. Lipid levels of family members should be tested.
§ Patients at high or moderately high risk will require simultaneous initiation of TLC and drug therapy.

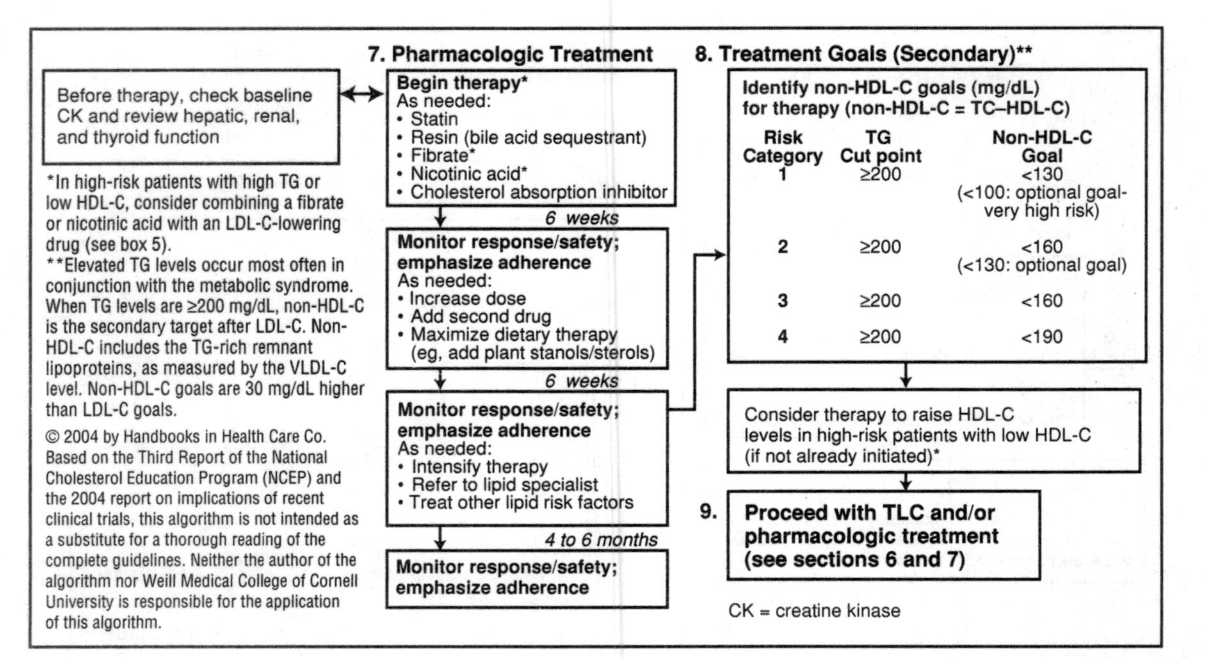
7. Pharmacologic Treatment
8. Treatment Goals (Secondary)**
Before therapy, check baseline CK and review hepatic, renal, and thyroid function
Begin therapy*
As needed:
• Statin
• Resin (bile acid sequestrant)
• Fibrate*
• Nicotinic acid*
• Cholesterol absorption inhibitor
6 weeks
Monitor response/safety; emphasize adherence
As needed:
• Increase dose
• Add second drug
• Maximize dietary therapy (eg, add plant stanols/sterols)
6 weeks
Monitor response/safety; emphasize adherence
As needed:
• Intensify therapy
• Refer to lipid specialist
• Treat other lipid risk factors
4 to 6 months
Monitor response/safety; emphasize adherence
Identify non-HDL-C goals (mg/dL) for therapy (non-HDL-C = TC–HDL-C)
Risk Category
TG Cut point
Non-HDL-C Goal
1
≥200
<130
(<100: optional goal-very high risk)
2
≥200
<160
(<130: optional goal)
3
≥200
<160
4
≥200
<190
Consider therapy to raise HDL-C levels in high-risk patients with low HDL-C (if not already initiated)*
9.
Proceed with TLC and/or pharmacologic treatment (see sections 6 and 7)
CK = creatine kinase
*In high-risk patients with high TG or low HDL-C, consider combining a fibrate or nicotinic acid with an LDL-C-lowering drug (see box 5).
**Elevated TG levels occur most often in conjunction with the metabolic syndrome. When TG levels are ≥200 mg/dL, non-HDL-C is the secondary target after LDL-C. Non-HDL-C includes the TG-rich remnant lipoproteins, as measured by the VLDL-C level. Non-HDL-C goals are 30 mg/dL higher than LDL-C goals.
© 2004 by Handbooks in Health Care Co.
Based on the Third Report of the National Cholesterol Education Program (NCEP) and the 2004 report on implications of recent clinical trials, this algorithm is not intended as a substitute for a thorough reading of the complete guidelines. Neither the author of the algorithm nor Weill Medical College of Cornell University is responsible for the application of this algorithm.

fore, waist circumference can have clinical significance. The ATP III guidelines define abdominal obesity as waist circumference >40 inches in men and >35 inches in women.

Based on information obtained in the medical interview and physical examination, the physician will be able to assign each patient to a CHD risk category (see Risk Classification). It is important to remember, however, that risk assessment and classification are often not straightforward—for example, consideration of information on the patient's medication history. In the ATP III guidelines, one of the major CHD risk factors is hypertension *or* use of antihypertensive medication. Therefore, even when blood pressure is <140/90 mm Hg (<130/<85 mm Hg for the metabolic syndrome), it is crucial to ascertain whether the patient is on antihypertensive therapy. Moreover, some medications may indicate the presence of an underlying condition that can cause secondary dyslipidemia, or a patient's dyslipidemia may be medication-related. In addition, physicians must remember that the CHD risk categories and the Framingham risk scoring system are tools whose use must be tempered by clinical judgment (see Clinical Judgment). There are many patients who are not easily categorized and whose actual risk may be far greater than their apparent risk (see The Metabolic Syndrome, below).

Laboratory Tests

Fasting lipoprotein profile. As described above, a fasting lipoprotein profile measures levels of TC, LDL-C, HDL-C, and TG. It is a fundamental component of risk assessment and the best tool for measuring patient compliance with lipid-regulating therapy. The fasting profile is preferable to a nonfasting measurement of TC and HDL-C. When TG concentrations are ≥200 mg/dL, the physician should also calculate the patient's level of non-HDL-C (TC – HDL-C). Non-HDL-C includes VLDL-cholesterol (VLDL-C), which is the most readily available measure of TG-rich remnant lipoproteins in clinical practice. In patients with elevated TG levels, non-HDL-C is a secondary target of risk-reduction therapy (Figure 3-1).

Table 3-1: Selected Causes of Primary Dyslipidemia

Hypercholesterolemia

Heterozygous familial hypercholesterolemia

- Autosomal dominant; LDL receptor defect; prevalence is 1 in 500 patients in Western Europe and North America; TC 350-500 mg/dL and increased risk for premature atherosclerosis; tendon xanthomata, corneal arcus

Homozygous familial hypercholesterolemia

- Autosomal dominant; LDL receptor defect; prevalence is 1 in 1,000,000 patients; TC 700-1,200 mg/dL and severe widespread premature atherosclerosis; cutaneous xanthomata, tendon xanthomata, corneal arcus

Familial defective apolipoprotein B-100

- Autosomal dominant; apo B mutation; prevalence is rare to 1 in 600 patients (varies by ethnicity); lipoprotein values and clinical features resemble heterozygous familial hypercholesterolemia

Polygenic hypercholesterolemia

- Various genetic defects; TC elevation less than heterozygous familial hypercholesterolemia

Disorders of HDL Metabolism

Familial hypoalphalipoproteinemia

- Familial low HDL-C; very rare; risk for premature coronary heart disease unclear (absent to increased)

Lecithin:cholesterol acyltransferase deficiency (partial)

- Fish-eye disease; HDL-C/apo A-I levels reduced, but premature atherosclerosis rare; corneal arcus

Familial apolipoprotein A-I/C-III deficiency

Apo A-I Milano

- Severe reductions in HDL cholesterol; no increase in CHD risk (found in a small northern Italian village)

ABC-A1 mutation (Tangier disease)

- Severe HDL-C deficiency; very rare; premature atherosclerosis

Primary Combined Hyperlipidemias

Familial combined hyperlipidemia

- Etiology unknown; overproduction of apo B in most cases; prevalence is 1 in 100 patients; TC 250-350 mg/dL and/or hypertriglyceridemia

Familial dysbetalipoproteinemia

- Other metabolic factors (eg, diabetes mellitus) usually required for full expression; prevalence is 1 in 5,000 patients; TC 300-600 mg/dL, TG 400-800 mg/dL or higher; palmar xanthomata; tuberoeruptive xanthomata

Primary Hypertriglyceridemia

Familial hypertriglyceridemia

- Mechanism unknown; prevalence is rare to 1 in 300 patients; TG 200-500 mg/dL to >1,000 mg/dL (risk for pancreatitis)

Familial chylomicronemia

- Autosomal recessive; lipoprotein lipase or apo C-II deficiency; rare

Lipoprotein lipase deficiency

Apolipoprotein C-III deficiency

HDL = high-density lipoprotein; LDL = low-density lipoprotein; TC = total cholesterol; TG = triglyceride

Adapted from *The ILIB Lipid Handbook for Clinical Practice*, 3rd ed. New York, International Lipid Information Bureau, 2003, pp 107-110.

Table 3-2: Selected Causes of Secondary Dyslipidemia

↑LDL-C	• Anorexia nervosa • Dysglobulinemia • Cholestasis • Hypothyroidism • Nephrotic syndrome	• Obstructive liver disease • Anabolic steroids • Corticosteroids • Cyclosporine • Progestins • Thiazide diuretics
↑TG	• Excess alcohol consumption • High-carbohydrate diet (>60% of caloric intake) • Obesity • Pregnancy • Cushing's syndrome • Diabetes mellitus • Hypothyroidism • Nephrotic syndrome • Chronic renal failure • Antiretroviral nucleoside analogues	• β-Blockers • Corticosteroids • Thiazide diuretics • Exogenous estrogens (oral administration) • Oral contraceptives • Retinoids • Tamoxifen
↓HDL-C	• Physical inactivity • Smoking • Diabetes mellitus • Obesity • Hypertriglyceridemia	• Anabolic steroids (oral administration) • β-Blockers • Progestins • Retinoids

Table 3-3: Life-Habit and Emerging Risk Factors

Life-Habit Risk Factors	Emerging Risk Factors
• Obesity • Physical inactivity • Atherogenic diet	• Lipoprotein(a) • Homocysteine • Proinflammatory factors (eg, C-reactive protein) • Prothrombotic factors (eg, fibrinogen) • Impaired fasting glucose • Subclinical atherosclerotic disease

Based on the premise that a VLDL-C level ≤30 mg/dL is normal, non-HDL-C goals are set at 30 mg/dL higher than LDL-C goals.

Fasting plasma glucose test. A fasting plasma glucose test is also important because it may signify the presence of insulin resistance or other risk factors for the metabolic syndrome.

Clinical chemistry. Tests of hepatic, renal, and thyroid function are necessary to evaluate whether a patient's dyslipidemia may be secondary to a disorder of the liver, kidneys, or thyroid. Tables 3-1 and 3-2 list selected causes of primary and secondary dyslipidemia. Any underlying disorders associated with secondary dyslipidemia must be treated before the physician can address the patient's lipid abnormalities. In addition, increased levels of lipoprotein (a) [Lp(a)] and homocysteine and the presence of proinflammatory (eg, C-reactive protein[8,9]) and prothrombotic markers may indicate elevated cardiovascular risk and help guide the intensity of risk-reduction therapy in selected patients[1] (see Appendix A, Case 6). These *emerging risk factors* (Table 3-3) are also implicated in the metabolic

Table 3-4: Types of Atherosclerotic Disease

CHD	Definite clinical and laboratory evidence of MI; clinically significant myocardial ischemia; history of coronary bypass surgery or coronary angioplasty; or angiogram demonstrating significant coronary atherosclerosis in the presence of clinical symptoms of CHD, although angiography is not recommended specifically to classify patients for cholesterol-lowering therapy.
PAD	Clinical signs and symptoms of ischemia of the extremities, accompanied by significant atherosclerosis on angiography or abnormalities of segment-to-arm pressure ratios or flow velocities.
AAA	An aneurysm in the abdominal aorta, usually found in an area with severe atherosclerosis.
Carotid	Cerebral symptoms (transient ischemic attacks or stroke) accompanied by ultrasound or angiographic evidence of significant atherosclerosis.

CHD = coronary heart disease; MI = myocardial infarction; PAD = peripheral arterial disease; AAA = abdominal aortic aneurysm

syndrome. Tests for the presence of emerging risk factors are recommended in (1) patients at moderately high or moderate risk (ie, ≥2 major risk factors and ≤20% risk), and (2) patients with 0 to 1 risk factor and an LDL-C level ≥190 mg/dL, presence of familial hypercholesterolemia, or a history of premature CHD in a first-degree relative.

Primary and Secondary Prevention

Traditionally, patients without clinical evidence of CHD were placed in a category of primary prevention, while those with known CHD were designated as candidates for secondary prevention.

However, the more recent emphasis on global risk assessment tends to blur the distinction between these two categories. For example, in the ATP III guidelines, the designation of high risk applies to patients with established CHD and also to those with CHD risk equivalents (ie, other forms of atherosclerotic disease [Table 3-4], diabetes, or ≥2 risk factors conferring a 10-year risk >20%). Treatment decisions for persons with CHD risk equivalents are based on the same criteria used for patients with established CHD. This has several implications, including an LDL-C goal <100 mg/dL. Results of recent clinical trials, which provide the rationale for a 2004 update of the ATP III guidelines, affirm the concept of treating high risk rather than just high cholesterol.[2]

Risk Classification

Risk Categories

Table 3-5 presents optimal, borderline, and high lipid levels (LDL-C, TC, and HDL-C) as classified by ATP III. Depending on each patient's level of risk for CHD, however, some variation in LDL-C targets is permissible. To determine target levels in individual cases, the following risk categories have been developed (Table 3-6):

- **High risk:** CHD or CHD risk equivalent
- **Moderately high risk:** ≥2 risk factors with a 10-year CHD risk of 10% to 20%
- **Moderate risk:** ≥2 risk factors with a 10-year risk <10%
- **Lower risk:** 0-1 risk factor (almost all people in this category have a 10-year CHD risk <10%)

For patients in the highest risk category, the recommended LDL-C target is <100 mg/dL. This goal is based on evidence from trials with both clinical and angiographic end points, as well as on data from prospective epidemiologic studies. For patients considered to be at very high risk, the 2004 ATP III update states that an LDL-C goal of <70 mg/dL is a therapeutic option based on clinical trial evidence. As discussed earlier, factors favoring this option are the presence of established CVD plus (1) multiple major risk factors, (2) severe or poorly controlled risk factors (especially cigarette smoking), (3) multiple risk factors of the metabolic syndrome (particularly TG level ≥200 mg/dL plus non-HDL-C ≥130 mg/dL with HDL-C <40 mg/dL), or (4) an acute coronary syndrome.

In the absence of CHD, other forms of atherosclerotic disease, or diabetes, 10-year risk must be calculated using the Framingham scoring system (Table 3-7). This system, which is straightforward and easy to use, assigns a point value to each of five risk factors: age, TC, smoking status, HDL-C, and systolic blood pressure. A table is then used to equate each patient's total number of points with a percent probability for developing CHD in the next 10 years.[1]

Although the Framingham score is an important tool, it may not adequately predict risk in certain patients (eg, those with a family history of premature CHD or CVD). Premature parental CHD (father <55 years, mother <65 years) is a major ATP III risk factor, but it is not part of the Framingham algorithm. This omission was based on doubt about the true magnitude of the associated independent risk, largely because of the retrospective or self-reported nature of the available data. Recently, however, an 8-year prospective Framingham Offspring Study with independent, uniform validation of events found that premature CVD in at least one parent was associated with increased cardiovascular risk in middle-aged men and women, particularly those at intermediate risk.[10] Following multivariate adjustment for traditional risk factors, premature parental CVD was associated with a significant twofold increase in cardiovascular risk in men and a nonsignificant 70% increase in women. This suggests that the addition of parental history may help refine risk assessment and guide treatment decisions in patients at intermediate risk based on other

Table 3-5: ATP III Classification of LDL-C, TC, and HDL-C Levels (mg/dL)

Lipid Level	ATP III Classification
LDL-C	
<100	Optimal
100-129	Near or above optimal
130-159	Borderline high
160-189	High
≥190	Very high
TC	
<200	Desirable
200-239	Borderline high
≥240	High
HDL-C	
<40	Low
≥60	High (desirable)

ATP = Adult Treatment Panel; LDL-C = low-density lipoprotein cholesterol; TC = total cholesterol; HDL-C = high-density lipoprotein cholesterol

Adapted from Executive Summary of the Third Report of the National Cholesterol Education Program (NCEP) Expert Panel on Detection, Evaluation, and Treatment of High Blood Cholesterol in Adults (Adult Treatment Panel III). *JAMA* 2001;285:2486-2497.

markers. The inclusion of validated parental CVD in a revised Framingham algorithm is being studied; in addition, certain novel or emerging risk factors, which may be more

Table 3-6: Categories of Risk That Modify LDL-C Targets

Risk Category	Definition
High Risk	
CHD and CHD risk equivalents	• Established CHD • CHD risk equivalent – Atherosclerotic disease - peripheral arterial disease - abdominal aortic aneurysm - carotid artery disease – Diabetes – ≥2 risk factors (10-yr risk >20%) - cigarette smoking - hypertension (≥140/90 mm Hg) low HDL-C (<40 mg/dL) - premature CHD in first-degree male (<55 yr) or female (<65 yr) relative - age (men ≥45 yr, women ≥55 yr)
Moderately High Risk	
≥2 risk factors	• 10-yr risk 10% to 20%*
Moderate Risk	
≥2 risk factors	• 10-yr risk <10%*
Lower Risk	
0-1 risk factor	• Most people with 0-1 risk factor have a 10-yr risk <10%; therefore, calculation of risk score is not necessary

CHD = coronary heart disease; CVD = cardiovascular disease; HDL-C = high-density lipoprotein cholesterol; LDL-C = low-density lipoprotein cholesterol; TC = total cholesterol; TG = triglyceride
Circulation 2004;110:227-239. *JAMA* 2001;285:2486-2497.

LDL-C Target
<100 mg/dL: strong recommendation <70 mg/dL: therapeutic option in very high-risk patients Factors favoring this option: established CVD plus • multiple major risk factors (especially diabetes) • severe/poorly controlled risk factors (especially cigarette smoking) • multiple risk factors of the metabolic syndrome (especially TG ≥200 mg/dL plus non-HDL-C ≥130 mg/dL with HDL-C <40 mg/dL) • an acute coronary syndrome
<130 mg/dL; <100 mg/dL therapeutic option**
<130 mg/dL
<160 mg/dL

*The percent probability of developing CHD is based on a risk scoring system derived from the Framingham Heart Study. The risk factors on which the Framingham score is based differ somewhat from the ATP III major risk factors, ie, premature familial CHD is not included, TC is included.
**Based on results of the Anglo-Scandinavian Cardiac Outcomes Trial—Lipid Lowering Arm (ASCOT-LLA). See Chapters 1 and 5.

Table 3-7: Framingham Point Scores Estimate of 10-Year Risk for Men

Age, years	Points
20-34	-9
35-39	-4
40-44	0
45-49	3
50-54	6
55-59	8
60-64	10
65-69	11
70-74	12
75-79	13

Total Cholesterol, mg/dL	Points				
	Age 20-39y	*Age 40-49y*	*Age 50-59y*	*Age 60-69y*	*Age 70-79y*
<160	0	0	0	0	0
160-199	4	3	2	1	0
200-239	7	5	3	1	0
240-279	9	6	4	2	1
≥280	11	8	5	3	1

	Points				
	Age 20-39y	*Age 40-49y*	*Age 50-59y*	*Age 60-69y*	*Age 70-79y*
Nonsmoker	0	0	0	0	0
Smoker	8	5	3	1	1

HDL-C, mg/dL	Points
≥60	-1
50-59	0
40-49	1
<40	2

Systolic BP, mm Hg	If Untreated	If Treated
<120	0	0
120-129	0	1
130-139	1	2
140-159	1	2
≥160	2	3

Point Total	10-Year Risk (%)
<0	<1
0	1
1	1
2	1
3	1
4	1
5	2
6	2
7	3
8	4
9	5
10	6
11	8
12	10
13	12
14	16
15	20
16	25
≥17	≥30

(continued on next page)

Table 3-7: Framingham Point Scores Estimate of 10-Year Risk for Women

Age, years	Points
20-34	-7
35-39	-3
40-44	0
45-49	3
50-54	6
55-59	8
60-64	10
65-69	12
70-74	14
75-79	16

Total Cholesterol, mg/dL	Points				
	Age 20-39y	*Age 40-49y*	*Age 50-59y*	*Age 60-69y*	*Age 70-79y*
<160	0	0	0	0	0
160-199	4	3	2	1	1
200-239	8	6	4	2	1
240-279	11	8	5	3	2
≥280	13	10	7	4	2

	Points				
	Age 20-39y	*Age 40-49y*	*Age 50-59y*	*Age 60-69y*	*Age 70-79y*
Nonsmoker	0	0	0	0	0
Smoker	9	7	4	2	1

HDL-C, mg/dL	Points
≥60	-1
50-59	0
40-49	1
<40	2

Systolic BP, mm Hg	If Untreated	If Treated
<120	0	0
120-129	1	3
130-139	2	4
140-159	3	5
≥160	4	6

Point Total	10-Year Risk (%)
<9	<1
9	1
10	1
11	1
12	1
13	2
14	2
15	3
16	4
17	5
18	6
19	8
20	11
21	14
22	17
23	22
24	27
≥25	≥30

Table 3-8: Risk Factors for the Metabolic Syndrome

Diagnosis of the metabolic syndrome is based on the presence of three or more of the following:

- Abdominal obesity (waist circumference >40 inches in men, >35 inches in women)*
- Triglycerides ≥150 mg/dL
- Low HDL-C (<40 mg/dL in men, <50 mg/dL in women)
- Blood pressure ≥130/≥85 mm Hg
- Fasting glucose ≥100** mg/dL

*ATP III recommends measuring waist circumference, rather than calculating body mass index (BMI), because abdominal obesity is more highly correlated with the metabolic risk factors than is an elevated BMI.
**Based on recent guidelines from the American Diabetes Association[11] and the National Heart, Lung, and Blood Institute/American Heart Association.[12]
HDL-C = high-density lipoprotein cholesterol

predictive than parental history, could one day play a role in risk estimation formulas.

For patients at moderately high or moderate risk, the target LDL-C level is <130 mg/dL. Those at lower risk have an LDL-C goal of <160 mg/dL.

Life-Habit and Emerging Risk Factors

The 2001 guidelines also identify *life-habit risk factors* and *emerging risk factors* that can affect a person's chance of developing CHD. These are not major independent risk factors, but, in selected patients, they can help guide a decision about the intensity of risk-reduction therapy. The life-habit risk factors (obesity, physical inactivity, and atherogenic diet) are targets for direct clinical intervention. The emerging risk factors are elevated Lp(a) and homocysteine levels, prothrombotic and proinflammatory factors, impaired fasting glucose, and evidence of subclinical atherosclerotic disease (Table 3-3).

The Metabolic Syndrome

The metabolic syndrome—a combination of major, life-habit, and emerging risk factors—can increase a patient's risk for CHD at any LDL-C level. Consequently, it is a potential secondary target of risk-reduction therapy, after LDL-C levels have been lowered. The metabolic syndrome is characterized by abdominal obesity, atherogenic dyslipidemia (elevated TG; small dense LDL particles; low HDL-C), elevated blood pressure, insulin resistance (with or without glucose intolerance), and prothrombic and proinflammatory factors. Table 3-8 presents the risk factors for clinical identification of the metabolic syndrome. The presence of three or more of these factors warrants a diagnosis of the metabolic syndrome.

Fredrickson Classification of the Hyperlipidemias

A useful model for classifying dyslipidemia is the Fredrickson scheme developed by investigators at the National Institutes of Health, in which the dyslipidemias are categorized based on elevations in TC and TG (Table 3-9).[13] Fredrickson phenotypes do not represent diagnoses because they do not address the primary or secondary origin of the dyslipidemia and do not account for HDL-C concentration.

The most atherogenic of the phenotypes are types IIa, IIb, and III, which are characterized by elevations in cholesterol alone, or in both cholesterol and TG. Types IV and V hyperlipidemias are less atherogenic, while type I, associated with chylomicronemia, has not been shown to be atherogenic.

Table 3-9: Fredrickson Classification of the Hyperlipidemias

Phenotype	Elevated Lipoprotein(s)	Elevated Lipid Levels
I	Chylomicrons	TG
IIa	LDL	TC
IIb	LDL and VLDL	TC, TG
III	IDL	TC, TG
IV	VLDL	TG, TC
V	VLDL and chylomicrons	TG, TC

IDL = intermediate-density lipoprotein; LDL = low-density lipoprotein; N = normal; TC = total cholesterol; TG = triglyceride; VLDL = very-low-density lipoprotein

However, clinicians should understand that, while chylomicronemia per se may not be atherogenic, chylomicron remnants are potentially atherogenic, and extreme elevations in TG carry a risk for pancreatitis.

Other Issues in Risk Assessment

Very High Levels of LDL-C

The presence of very high levels of LDL-C usually signifies a genetic cause (eg, monogenic familial hypercholesterolemia, familial defective apolipoprotein [apo] B, polygenic hypercholesterolemia). Case 5 in Appendix A presents such a patient and highlights the importance of cholesterol testing beginning at age 20, as recommended by the ATP III guidelines, to prevent the development of premature CHD. When very high LDL-C levels are detected, it is important to assess cholesterol levels in the patient's family members as well.

Plasma TC	Plasma TG	Relative Frequency (%)*
N to ↑	↑↑↑↑	<1
↑↑	N	10
↑↑	↑↑	40
↑↑	↑↑↑	<1
N to ↑	↑↑	45
↑ to ↑↑	↑↑↑↑	5

*Approximate % of patients in the United States with hyperlipidemia (Adapted from *The ILIB Lipid Handbook for Clinical Practice*, 2nd ed. New York, International Lipid Information Bureau 2000, p 45.)

Hypertriglyceridemia

The role of hypertriglyceridemia as a cardiovascular risk factor is controversial. According to prospective observational studies, including a meta-analysis of 17 population-based trials, elevated TG levels are a risk factor for CVD independent of HDL-C levels.[14-16] However, some epidemiologic studies have found that the predictive value of hypertriglyceridemia loses significance after adjustment for covariates (eg, HDL-C, plasma glucose levels).[17,18] Therefore, it is unclear whether hypertriglyceridemia is directly atherogenic or whether it is atherogenic through its link to abnormalities of the metabolic syndrome[19,20] (see Chapter 1, Refining Treatment Guidelines).

Factors that can contribute to hypertriglyceridemia include obesity (also characteristic of the metabolic syndrome), excess alcohol intake, a high-carbohydrate diet (>60% of calories), underlying disease states (eg, diabetes mellitus, renal failure, nephrotic syndrome), certain drugs (eg, corticosteroids, estrogens, retinoids, higher doses of β-adrenergic blockers; see Table 3-2), and genetic disorders (eg, familial combined hyperlipidemia, familial hypertriglyceridemia, familial dysbetalipoproteinemia). The association between renal abnormalities and hypertriglyceridemia reinforces the importance of obtaining laboratory values for renal function when assessing patient risk.

Evidence suggests that some TG-rich lipoproteins may be atherogenic (eg, partially degraded VLDL, commonly termed remnant lipoproteins).[1,21,22] The role of TG-rich lipoproteins in atherogenesis is addressed in more detail in Chapter 2. In the clinical setting, VLDL-C is the most easily available measure of TG-rich remnant lipoproteins. Therefore, non-HDL-C (TC – HDL-C), which includes VLDL-C, may be a secondary target of lipid lowering in patients with high TG levels (200 to 499 mg/dL).

Low Levels of HDL-C

Low HDL-C (ie, <40 mg/dL) is a strong independent predictor of CHD.[1,23] In the secondary-prevention Veterans Affairs High-Density Lipoprotein Trial (VA-HIT),[23] which enrolled 2,531 men whose primary lipid abnormality was low HDL-C,[24] gemfibrozil (Lopid®) raised HDL-C levels by 6%, lowered TG concentrations by 31%, and reduced the risk for a composite primary end-point event (ie, nonfatal myocardial infarction [MI] or death from CHD) by 22% compared with placebo (see Chapter 1, Refining Treatment Guidelines). There was no change in LDL-C levels.[24] Data from a multivariate analysis indicate that only on-treatment HDL-C levels predicted coronary artery disease events at 1 year.[25] These outcomes from VA-HIT are consistent with the results of earlier studies (eg, the Framingham Heart Study, the Helsinki Heart Study), in which HDL-C was shown to be an independent risk factor for CVD.[23]

The Air Force/Texas Coronary Atherosclerosis Prevention Study (AFCAPS/TexCAPS), a primary-prevention trial involving 6,605 men and women, found that lovastatin (Mevacor®)-treated patients in the lowest tertile of baseline HDL-C experienced a decrease of about 45% in risk for a primary end-point event (ie, fatal or nonfatal MI, unstable angina, or sudden cardiac death) compared with a 37% risk reduction in the overall study sample.[23,26]

In AFCAPS/TexCAPS, baseline concentrations of HDL-C and apo A-I, the principal apolipoprotein component of HDL, were significantly associated with the occurrence of a first acute major coronary event.[27] There were also significant associations between a primary outcome event and baseline ratios of LDL-C/HDL-C and TC/HDL-C. Although low-density lipoprotein cholesterol and TC did not reach statistical significance as predictors of risk, either at baseline or at year 1, levels of apo B (the principal apolipoprotein component of LDL) and the apo B/apo A-I ratio at baseline were significant risk factors. At year 1, only apo A-I, apo B, and the ratio of apo B/apo A-I were significant predictors of risk. Based on multivariate analysis, apo A-I was the only significant risk determinant at baseline, at 1-year follow-up, and for the percent change between baseline and follow-up.[27] These data suggest that the benefit of lovastatin may in part be a function of changes in apo B and apo A-I.[27]

The importance of HDL-C in assessing cardiovascular risk was reaffirmed by the results of the Quebec Cardiovascular Study, a prospective trial involving 2,103 men followed for a 5-year period.[28] In this study, HDL-C was found to be an independent predictor of a first ischemic event (ie, typical effort angina, coronary insufficiency, nonfatal MI, or coronary death). Other independent predictors included LDL-C, diabetes, and systolic blood pressure. However, the investigators also found that the cholesterol/HDL-C ratio was the single best lipid predictor of ischemic heart disease. Subjects in the lowest quartile of HDL-C had a 156-fold increase in the odds of having a cholesterol/HDL-C ratio >6 compared with those in the upper quartile, whereas the odds for subjects in the highest quartile of LDL-C were 14.4 times greater than for those in the lowest quartile. These results suggest that the "proper evaluation and optimal management" of coronary risk extends beyond LDL-C, and that raising HDL-C may help reduce risk as part of a global risk management strategy.[28]

According to the ATP III guidelines, low HDL-C is both a major risk factor for CHD and one of the factors used in the Framingham scoring system to estimate 10-year risk. It is also one of the risk determinants for the metabolic syndrome. In ATP III, an HDL-C concentration ≥60 mg/dL is considered to be a 'negative' risk factor (ie, its presence reduces a patient's total number of risk factors by one).

Clinical Judgment

Clinical judgment can be critical in evaluating a patient's risk for CHD. Although the ATP III guidelines present a detailed system for risk assessment, they acknowledge that certain cases cannot be strictly categorized. Case 2 in Appendix A is such an example. According to ATP III, this patient has two of the major CHD risk factors: HDL-C <40 mg/dL and premature CHD in a first-degree family member. However, premature familial CHD is not one of the risk factors in the Framingham scoring system. Therefore, according to the Framingham criteria, this patient has a 10-year risk <10%. However, the physician must regard this patient as being at considerably greater risk. In addition to having a father who underwent coronary artery bypass graft (CABG) surgery after an MI at 53 years of age, the patient has four of the risk factors for the metabolic syndrome. Although his systolic blood pressure (136 mm Hg) is below the major risk factor cut point of 140 mm Hg, it is elevated in the context of the metabolic syndrome. The type of treatment warranted by a risk <10% would not be appropriate in this case.

Case 5 also illustrates the importance of clinical judgment. This 22-year-old woman has one major risk factor: a father who had an MI at 50 years of age. She is healthy, exercises regularly, and is a vegetarian. Because her HDL-C (84 mg/dL) is well above the major risk factor cut point of 40 mg/dL, it is regarded as a 'negative' risk factor. According to this guideline, the patient would then have no major risk factors. However, the patient's LDL-C level is 208 mg/dL and her TG concentration is 185 mg/dL. According to ATP III guidelines, patients with 0 to 1 risk factor should be considered for lipid-lowering drug therapy if their LDL-C concentration is ≥190 mg/dL. In such a case, statin therapy may be the most appropriate option. Although this patient's LDL-C concentration exceeds the specified level, she is a woman of childbearing age. Because pregnancy and lactation are contraindications for statin therapy, the physician must exercise clinical judgment in determining whether to prescribe a drug in this class. If it is prescribed, the patient must be cautioned to use a contraceptive and to discontinue statin treatment before becoming pregnant. Prescribing a statin requires a reasonable degree of confidence that the patient will adhere to this warning. As previously stated, the medical interview provides the physician with the opportunity to learn about patient beliefs that can influence compliance with treatment.

Global Risk

In estimating CHD risk, the synergistic effects of multiple risk factors have become more widely recognized. The ATP guidelines represent an important step toward developing a risk assessment strategy based on multiple lipid and nonlipid factors (Table 3-10). Groups such as the American Heart Association (AHA) and the American College of Cardiology have issued recommendations on managing post-MI patients that propose a multipart strategy involving not only treating dyslipidemia, but also managing hypertension and encouraging patients to stop smoking. Recent clinical trials (see Chapter 1) provide strong evidence to support the concept of treating high risk, not just high cholesterol.

Physician Compliance: Issues

Lipid testing. Appropriate assessment of each patient's cardiovascular risk is the basis for treatment decisions and

Table 3-10: Components of Risk Assessment

- Fasting lipoprotein profile
  - TC level
  - LDL-C level
  - HDL-C level
  - TG level
- Major risk factors
- Risk classification
  - established CHD
  - CHD risk equivalents
  - Framingham risk scoring system
- The metabolic syndrome
  - life-habit risk factors
  - emerging risk factors
- Clinical judgment

TC = total cholesterol; LDL-C = low-density lipoprotein cholesterol; HDL-C = high-density lipoprotein cholesterol; TG = triglycerides; CHD = coronary heart disease

can have a profound effect on the success or failure of prevention strategies. Nevertheless, physicians do not always follow the fundamental guidelines for risk assessment, both in community-based settings and in major health-care institutions (Table 3-11). Investigators at the Mount Sinai Medical Center in New York conducted a retrospective cross-sectional study involving two practice settings: a cardiology clinic (131 patients) and a cardiology private practice (139 patients).[29] These practice settings represented patients of differing socioeconomic groups, as reflected in financial reimbursement for services (eg, percentage reimbursed by Medicare). A systematic review of the 270 medical charts at

Table 3-11: Issues Concerning Physician Compliance With Risk Assessment Guidelines

- Understanding the importance of risk assessment can provide the basis for all future therapeutic decisions; interventions can reduce the risk for coronary heart disease by approximately 30%.
- Understanding the needs of special populations
  - to determine risk in women (ie, cardiovascular risk increases after menopause; diabetes abolishes the gender difference of reduced cardiovascular risk in premenopausal women)
  - to determine risk in the elderly (ie, emphasis on reducing morbidity rather than mortality)
- Developing organizational systems to facilitate adherence to risk assessment guidelines

both settings showed great variation in physician performance. The proportion of patients per physician who had their lipid levels measured ranged from 0% to 83%, with no statistically significant difference between settings in the overall percentage of patients whose lipid values were tested (40% in the clinic vs 47% in the private office setting). Based on these findings and other studies, the authors concluded that individual physicians, rather than the patient's socioeconomic status, accounted for variations in the management of lipid disorders. These results were surprising, since the investigators had surmised that differences in lipid management would be attributable to patient characteristics (eg, inability to follow instructions correctly because of inadequate English or lack of education) or to a more aggressive approach on the part of physicians whose patients had higher levels of education.[29]

Other data also indicate that clinicians do not always adhere to the basic requirements of risk assessment. In one survey conducted at a community-based family practice residency program, English-speaking Medicaid patients were less likely than patients with private insurance to have received cholesterol screening (39% vs 65% screened) over a 5-year period.[30] In comparison, Medicaid and private insurance patients were screened with equal frequency for hypertension and cervical cancer. Because access to testing is the same or better with Medicaid compared with private insurance, the gap in cholesterol screening may be partly explained by differences in the attention that physicians pay to this aspect of care. Studies have also found that cholesterol levels often are not measured in patients hospitalized for an acute MI[31] and in postsurgical cardiac patients.[32] According to an observational study, for example, just 24% of patients with acute MI underwent cholesterol screening in Worcester, Massachusetts, metropolitan area hospitals in 1997. The same study, which surveyed seven 1-year periods between 1986 and 1997, found that rates of cholesterol screening in such patients may, in fact, be declining.[31] High-density lipoprotein cholesterol, in particular, is frequently not monitored in coronary artery disease and postsurgical cardiac patients.[32]

These results are corroborated by a study that examined many preventive services, including cholesterol screening, in 44 primary care practices in the Midwest.[33] The rate of cholesterol screening ranged from 45% to 88%. In contrast, the percentage of patients who received a Papanicolaou (Pap) smear for cervical cancer ranged from 70% to 93%. The authors of this study suggest that the wide variation in rates of preventive services, both within and between clinics, may be explained by inadequate organizational systems.

At-risk patients: testing family members. Another important physician compliance issue concerns the screening of first-degree relatives of patients with CVD, as recommended by the 1993 NCEP guidelines. The 1996-1997 American College of Cardiology Evaluation of Preventive Therapeutics (ACCEPT) study was a nationwide survey of 5,553 patients admitted to 53 hospitals for first bypass surgery, first angioplasty, an acute MI, or myocardial ischemia.[34] The survey found that less than 1% of inpatient medical records included a discharge recommendation to screen family members of patients younger than 55 years of age. Six-month follow-up interviews revealed that just 17.8% of these patients had their family members screened after the cardiovascular event, and only 19.6% of patients with a recognized family history of premature coronary artery disease had their family members evaluated. These results suggest that physicians do not comply with national recommendations for screening the family members of high-risk patients.

Awareness of risk in women. Although CVD (including heart disease) is the leading cause of death in women,[35] physicians may have a misconception about women's risk for CHD.[36] This may explain the results of a national telephone survey conducted by researchers at the Columbia University College of Physicians and Surgeons in New York City. Despite the fact that 86% of the 1,002 women surveyed said that they see a physician for regular checkups, more than half of this group stated that their physician had never discussed CVD.[36] This included 47% of women aged 45 to 59 years and 44% of those older than 60 years. Of the women who regularly see a physician, 98% and 97% reported having their blood pressure and weight checked, respectively, but only 13% indicated that their waist/hip ratio was calculated, and just 50% stated that their physician recommended a cholesterol test. The investigators concluded that both the extent and the quality of cardiovascular risk assessment in women may be questionable. This study also provides insight into women's self-perception and the potential role of the physician in patient education. Although 74% of the women surveyed considered themselves knowledgeable about women's health issues, 44% stated that they were unlikely to have a heart attack and 58% felt that they were more likely to die of breast cancer than of CHD. According to the AHA, CVD (including heart disease) is the cause of death in 1 of every 2.5 women each year, whereas 1 in 30 women dies of breast cancer.[33] Chapter 6 examines CHD prevention in women.

Communication skills. Physician communication skills are fundamental to effective patient risk assessment. The history-taking interview is an opportunity for physicians to offer emotional support, educate patient and family about CVD and preventive measures, determine whether to assess CHD risk in family members, and involve patient and family members as informed participants in preventive care.[37,38] It has been shown that these critical goals can be achieved without sacrificing efficiency or extending the length of the office visit.[38]

Complexity of the problem. Three important physician-related reasons may explain inadequate screening for cardiovascular risk: (1) insufficient attention to this aspect of care, (2) inadequate organizational systems, and (3) misperception about the risk for CVD in women (Table 3-10). The reasons for physician inattention are numerous, complex, and beyond the scope of this chapter. Some of these reasons, which can extend beyond risk assessment to other aspects of preventive cardiology, are explained in Chapter 4 (see Physician Compliance). Figure 3-1 presents the essential steps in risk assessment. We hope that the information in this book will encourage physicians to comply more consistently with the risk assessment procedures recommended by the ATP III guidelines. Such compliance can lead to improved cardiovascular care for our patients.

References

1. Executive Summary of the Third Report of the National Cholesterol Education Program (NCEP) Expert Panel on Detection, Evaluation, and Treatment of High Blood Cholesterol in Adults (Adult Treatment Panel III). *JAMA* 2001;285:2486-2497.

2. Grundy AM, Cleeman JI, Merz CNB, et al: Implications of recent clinical trials for the National Cholesterol Education Program Adult Treatment Panel III guidelines. *Circulation* 2004;110:227.

3. Summary of the Second Report of the National Cholesterol Education Program (NCEP) Expert Panel on Detection, Evaluation, and Treatment of High Blood Cholesterol in Adults (Adult Treatment Panel II). *JAMA* 1993;269:3015-3023.

4. Framingham Heart Study: 50 years of research success. National Heart, Lung, and Blood Institute. National Institutes of Health. Available at http://www.nhlbi.nih.gov/about/framingham/index.html. Accessed September 22, 2002.

5. Frankel RM, Stein T: Getting the most out of the clinical encounter: the four habits model. *J Med Pract Manage* 2001;16:184-191.

6. Svensson S, Kjellgren KI, Ahlner J, et al: Reasons for adherence with antihypertensive medication. *Int J Cardiol* 2000;76:157-163.

7. Kiortsis DN, Giral P, Bruckert E, et al: Factors associated with low compliance with lipid-lowering drugs in hyperlipidemic patients. *J Clin Pharm Ther* 2000;25:445-451.

8. Ridker PM, Hennekens CH, Buring JE, et al: C-reactive protein and other markers of inflammation in the prediction of cardiovascular disease in women. *N Engl J Med* 2000;342:836-843.

9. Ridker PM, Cushman M, Stampfer MJ, et al: Inflammation, aspirin, and the risk of cardiovascular disease in apparently healthy men. *N Engl J Med* 1997;336:973-979.

10. Lloyd-Jones DM, Nam BH, D'Agostino RB Sr, et al: Parental cardiovascular disease as a risk factor for cardiovascular disease in middle-aged adults: a prospective study of parents and offspring. *JAMA* 2004;291:2204-2211.

11. Follow-up report on the diagnosis of diabetes mellitus. The Expert Committee on the Diagnosis and Classification of Diabetes Mellitus. *Diabetes Care* 2003;26:3160-3167.

12. Grundy SM, Brewer HB, Cleeman JI, et al, for the Conference participants: Definition of metabolic syndrome. Report of the National Heart, Lung, and Blood Institute/American Heart Association Conference on Scientific Issues Related to Definition. *Circulation* 2004;109:433-438.

13. Fredrickson DS, Levy RI, Lees RS: Fat transport in lipoproteins—an integrated approach to mechanisms and disorders. *N Engl J Med* 1967;276:34-43 contd.

14. Hokanson JE, Austin MA: Plasma triglyceride level is a risk factor for cardiovascular disease independent of high-density lipo-

protein cholesterol level: a meta-analysis of population-based prospective studies. *J Cardiovasc Risk* 1996;3:213-219.

15. Jeppesen J, Hein HO, Suadicani P, et al: Triglyceride concentration and ischemic heart disease: an eight-year follow-up in the Copenhagen Male Study. *Circulation* 1998;97:1029-1036.

16. Assmann G, Schulte H, von Eckardstein A: Hypertriglyceridemia and elevated lipoprotein(a) are risk factors for major coronary events in middle-aged men. *Am J Cardiol* 1996;77:1179-1184.

17. Criqui MH, Heiss G, Cohn R, et al: Plasma triglyceride level and mortality from coronary heart disease. *N Engl J Med* 1993;328:1220-1225.

18. Castelli WP: The triglyceride issue: a view from Framingham. *Am Heart J* 1986;112:432-437.

19. Grundy SM, Vega GL: Two different views of the relationship of hypertriglyceridemia to coronary heart disease. Implications for treatment. *Arch Intern Med* 1992;152:28-34.

20. Cullen P: Evidence that triglycerides are an independent coronary heart disease risk factor. *Am J Cardiol* 2000;86:943-949.

21. Cohn JS, Marcoux C, Davignon J: Detection, quantification, and characterization of potentially atherogenic triglyceride-rich remnant lipoproteins. *Arterioscler Thromb Vasc Biol* 1999;19:2474-2486.

22. Krauss RM: Atherogenicity of triglyceride-rich lipoproteins. *Am J Cardiol* 1998;81(4A):13B-17B.

23. Boden WE: High-density lipoprotein cholesterol as an independent risk factor in cardiovascular disease: assessing the data from Framingham to the Veterans Affairs High-Density Lipoprotein Intervention Trial. *Am J Cardiol* 2000;85(suppl):19L-22L.

24. Rubins HB, Robins SJ: Conclusions from the VA-HIT study. *Am J Cardiol* 2000;86:543-544.

25. Robins SJ, Collins D, Wittes JT, et al: Relation of gemfibrozil treatment and lipid levels with major coronary events: VA-HIT: a randomized controlled trial. *JAMA* 2001;285:1585-1591.

26. Downs JR, Clearfield M, Weis S, et al: Primary prevention of acute coronary events with lovastatin in men and women with average cholesterol levels: results of AFCAPS/TexCAPS. Air Force/ Texas Coronary Atherosclerosis Prevention Study. *JAMA* 1998;279:1615-1622.

27. Gotto AM Jr, Whitney E, Stein EA, et al: Relation between baseline and on-treatment lipid parameters and first acute major coronary events in the Air Force/Texas Coronary Atherosclerosis Prevention Study (AFCAPS/TexCAPS). *Circulation* 2000;101:477-484.

28. Després JP, Lemieux I, Dagenais GR, et al: HDL-cholesterol as a marker of coronary heart disease risk: the Québec cardiovascular study. *Atherosclerosis* 2000;153:263-272.

29. Harnick DJ, Cohen JL, Schechter CB, et al: Effects of practice setting on quality of lipid-lowering management in patients with coronary artery disease. *Am J Cardiol* 1998;81:1416-1420.

30. Hueston WJ, Spencer E, Kuehn R: Differences in the frequency of cholesterol screening in patients with Medicaid compared with private insurance. *Arch Fam Med* 1995;4:331-334.

31. Yarzebski J, Spencer F, Goldberg RJ, et al: Temporal trends (1986-1997) in cholesterol level assessment and management practices in patients with acute myocardial infarction: a population-based perspective. *Arch Intern Med* 2001;161:1521-1528.

32. Smith SC Jr: Clinical treatment of dyslipidemia: practice patterns and missed opportunities. *Am J Cardiol* 2000;86(12A):62L-65L.

33. Solberg LI, Kottke TE, Brekke ML: Variation in clinical preventive services. *Eff Clin Pract* 2001;4:121-126.

34. Swanson JR, Pearson TA: Screening family members at high risk for coronary disease. Why isn't it done? *Am J Prev Med* 2001;20:50-55.

35. American Heart Association: Heart disease and stroke statistics—2004 update. Available at: http://www.americanheart.org/presenter.jhtml?identifier=3000090. Accessed September 30, 2004.

36. Legato MJ, Padus E, Slaughter E: Women's perceptions of their general health, with special reference to their risk of coronary artery disease: results of a national telephone survey. *J Womens Health* 1997;6:189-198.

37. Dube CE, O'Donnell JF, Novack DH: Communication skills for preventive interventions. *Acad Med* 2000;75(7 suppl):S45-S54.

38. Marvel MK, Doherty WJ, Weiner E: Medical interviewing by exemplary family physicians. *J Fam Pract* 1998;47:343-348.

Therapeutic Options: Dietary and Other Nondrug Interventions

The goal of dietary intervention is to reduce a patient's risk for coronary heart disease (CHD) by modifying risk factors that have been linked to food intake. This chapter will explain the dietary guidelines issued by the American Heart Association (AHA)[1] and by the National Cholesterol Education Program (NCEP) Expert Panel on Detection, Evaluation, and Treatment of High Blood Cholesterol in Adults (Adult Treatment Panel III [ATP III]).[2,3] Both sets of recommendations also stress the importance of physical activity.

Dietary practices can modify cardiovascular risk by helping to achieve and maintain desirable blood levels of total cholesterol (TC) and low-density lipoprotein cholesterol (LDL-C), a healthy body weight, and normal blood pressure.[1] Because a high-fat diet may offset cholesterol reductions achieved with drug therapy, patients must be encouraged to maintain dietary therapy even when taking a lipid-modifying agent. For those who have trouble adhering to nutritional recommendations, registered dietitians can play a critical role. Both the AHA and the NCEP use the term *medical nutrition therapy* to denote the more intensive guidance and supervision provided by qualified dietitians and nutritionists. To promote compliance, it is also important for the physician to maintain enthusiasm and support for each patient's efforts at dietary change.

Dietary modification combined with increased physical activity can also raise high-density lipoprotein cholesterol (HDL-C) levels and decrease the risk for diabetes. Exercise, together with caloric restriction, is particularly important for obese patients. All exercise programs should correspond to the patient's degree of fitness, cardiac status, and personal interests. Aerobic activity that stimulates the cardiovascular system should be emphasized.

The dietary recommendations in the AHA and NCEP guidelines are revisions of the Step I and Step II programs that both groups previously advocated for primary and secondary prevention.

In addition to its general dietary guidelines published in 2000, the AHA has issued a series of scientific reports on individual diet- and lifestyle-related topics. In large part, this chapter will discuss the dietary guidelines. A selected list of other AHA reports can be found in Table 4-1.

Following an examination of the AHA recommendations, this chapter describes the nutritional guidelines of the NCEP in more detail. The chapter concludes with information to help physicians implement the guidelines in their daily practice.

American Heart Association Guidelines

In the revised AHA guidelines, the Step I program is superseded by recommendations for the general population; in place of the Step II diet, the AHA advocates medical nutrition therapy for high-risk subgroups.

While emphasizing the need to maintain healthy dietary practices throughout life, the current AHA guidelines are designed as a framework within which there is room for adjustments based on individual health status, food preferences, and cultural background. The AHA guidelines are based on achieving the following objectives, which are regarded as being of equal importance in reducing cardiovascular risk: a

Table 4-1: American Heart Association: Statements on Diet and Lifestyle*

- Grundy SM, Hansen B, Smith SC Jr, et al: Clinical management of metabolic syndrome: report of the American Heart Association/National Heart, Lung, and Blood Institute/American Diabetes Association Conference on Scientific Issues Related to Management. *Circulation* 2004;109:551-556.
- Kavey RE, Daniels SR, Lauer RM, et al: American Heart Association guidelines for primary prevention of atherosclerotic cardiovascular disease beginning in childhood. *Circulation* 2003;107:1562-1566.
- Krauss RM, Eckel RH, Howard B, et al: AHA dietary guidelines. Revision 2000: a statement for healthcare professionals from the Nutrition Committee of the American Heart Association. *Circulation* 2000;102:2284-2299.
- Kris-Etherton P, Daniels SR, Eckel RH, et al: Summary of the Scientific Conference on Dietary Fatty Acids and Cardiovascular Health: conference summary from the Nutrition Committee of the American Heart Association. *Circulation* 2001;103:1034-1039.
- Kris-Etherton PM, Harris WS, Appel LJ: Fish consumption, fish oil, omega-3 fatty acids, and cardiovascular disease. *Circulation* 2002;106:2747-2757. Correction: *Circulation* 2003;107:512.
- Kris-Etherton P, Eckel RH, Howard BV, et al: AHA Science Advisory. Lyon Diet Heart Study. Benefits of a Mediterranean-style, National Cholesterol Education Program/American Heart Association Step I dietary pattern on cardiovascular disease. *Circulation* 2001;103:1823-1825.
- Lichtenstein AH, Deckelbaum RJ: AHA Science Advisory. Stanol/sterol ester-containing foods and blood cholesterol levels. A statement for healthcare professionals from the Nutrition Committee of the Council on Nutrition, Physical Activity, and Metabolism of the American Heart Association. *Circulation* 2001;103:1177-1179.

- Lichtenstein AH, Van Horn L: Very low fat diets. *Circulation* 1998;98:935-939.
- Ockene IS, Miller NH: Cigarette smoking, cardiovascular disease, and stroke. A statement for healthcare professionals from the American Heart Association. American Heart Association Task Force on Risk Reduction. *Circulation* 1997;96:3243-3247.
- Pollock ML, Franklin BA, Balady GJ, et al: AHA Science Advisory. Resistance exercise in individuals with and without cardiovascular disease: benefits, rationale, safety, and prescription. An advisory from the Committee on Exercise, Rehabilitation, and Prevention, Council on Clinical Cardiology, American Heart Association; Position paper endorsed by the American College of Sports Medicine. *Circulation* 2000;101:828-833.
- Steinberger J, Daniels SR: Obesity, insulin resistance, diabetes, and cardiovascular risk in children: an American Heart Association scientific statement from the Atherosclerosis, Hypertension, and Obesity in the Young Committee (Council on Cardiovascular Disease in the Young) and the Diabetes Committee (Council on Nutrition, Physical Activity, and Metabolism). *Circulation* 2003;107:1448-1453.
- Thompson PD, Buchner D, Pina IL, et al: Exercise and physical activity in the prevention and treatment of atherosclerotic cardiovascular disease: a statement from the Council on Clinical Cardiology (Subcommittee on Exercise, Rehabilitation and Prevention) and the Council on Nutrition, Physical Activity, and Metabolism (Subcommittee on Physical Activity). *Circulation* 2003;107:3109-3116.

*A complete list of American Heart Association position papers and guidelines on lifestyle-related and other topics is available at http://myamericanheart.org/portal/professional/guidelines.

Table 4-2: Comparison Between AHA and ATP III Guidelines

	AHA
Primary goals of dietary/lifestyle therapy	Achieve: • healthy body weight • desirable blood cholesterol/ lipoprotein profile • desirable blood pressure
Secondary goal of therapy (in moderate- and lower-risk patients)	
Patient classification	• General population • High-risk groups: – elevated LDL-C levels or preexisting CVD – diabetes or insulin resistance – congestive heart failure – kidney disease
Risk stratification	• No further risk stratification • Research recommended into role of specific foods/ nutrients/genetic factors
Medical nutrition therapy	• Recommended for patients in the four high-risk groups

AHA = American Heart Association; ATP = Adult Treatment Panel; CHD = coronary heart disease; CVD = cardiovascular disease; LDL-C = low-density lipoprotein cholesterol; VLDL-C = very low-density lipoprotein cholesterol

ATP III

- Lower LDL-C levels*
- Modify lifestyle-related risk factors** (regardless of LDL-C levels in high- or moderately high-risk patients)

- Lower non-HDL-C level***
- Treat the metabolic syndrome by weight loss/intensified physical activity

- CHD/CHD risk equivalents (including ≥2 risk factors with 10-yr CHD risk >20%)
- Moderately high risk: ≥2 risk factors (10-yr CHD risk 10% to 20%)
- Moderate risk: ≥2 risk factors with 10-yr risk <10%
- Low risk: 0-1 risk factor (almost all people in this category have a risk <10%; 10-yr risk assessment not necessary)

- Further risk stratification according to:
  - lifestyle-related risk factors**
  - emerging risk factors (lipoprotein[a], homocysteine, prothrombotic/proinflammatory factors, fasting glucose, subclinical atherosclerotic disease)

- Advisable for any patient, at discretion of physician

*In high-risk patients with TG elevations or low HDL-C, consider combining a fibrate or nicotinic acid with LDL-lowering therapy.
**Lifestyle-related risk factors include obesity, physical inactivity, elevated triglycerides, low HDL-C, or the metabolic syndrome.
***Two pharmacologic approaches: intensify LDL-lowering therapy or further decrease VLDL-C by cautiously adding a fibrate or nicotinic acid (if not already combined with LDL-lowering therapy; see note concerning high-risk patients).

(continued on next page)

Table 4-2: Comparison Between AHA and ATP III Guidelines *(continued)*

	AHA
Desirable lipid levels	• Not specified • Recommendation to follow NCEP/ATP guidelines
HDL-C levels	• High HDL-C is inversely associated with CHD. • Low HDL-C is associated with other CHD risk factors (eg, adiposity, sedentary lifestyle) that require treatment. • Insufficient evidence that raising HDL-C reduces CHD risk.

AHA = American Heart Association; ATP = Adult Treatment Panel; CHD = coronary heart disease; HDL-C = high-density lipoprotein cholesterol; LDL-C = low-density lipoprotein cholesterol; TG = triglyceride

healthy body weight; a desirable blood cholesterol and lipoprotein profile; and a desirable blood pressure (Table 4-2).

These broad principles are intended for the general population older than 2 years. In addition, the AHA identifies four groups that are at increased risk for cardiovascular disease (CVD) or a coronary event. Persons in these groups should receive more-intensive medical nutrition therapy (see below).

Although recent research has established that a regimen low in saturated fats is safe for children,[4-6] it cannot be as-

ATP III

- Specified

- Low HDL-C is a strong independent predictor of CHD.
- Low HDL-C is a risk factor used to modify LDL-C goal and estimate 10-year CHD risk.
- There is insufficient evidence to support a numeric goal for raising HDL-C.
- In high-risk patients with low HDL-C, initiation of a fibrate or nicotinic acid with LDL-lowering therapy can be considered.
- Low HDL-C is a feature of the metabolic syndrome and is associated with other CHD risk factors (eg, obesity, high TG, insulin resistance, diabetes) that require treatment.

sumed that a diet suitable for adults is also nutritionally appropriate for the pediatric population. Therefore, care must be taken to ensure nutrition that is sufficient for normal growth and development. Additional studies are needed to understand the relationship between a healthy diet and physical activity in childhood and the prevention of CVD later in life. Other areas for future investigation include the effects of nutritional intervention on obesity, type 2 diabetes, elevated LDL-C and triglyceride (TG) levels, low HDL-C concentrations in children and adolescents, and genetic factors that influence individual responses to nutrition.

Dietary Recommendations

The AHA guidelines provide specific dietary recommendations for the general population to prevent or delay the development of CVD. In addition to nutritional content, portion number and size are important in maintaining a balance between calorie intake and energy needs.

Dietary fats should constitute no more than 30% of total daily intake (Table 4-3). Within this limit, saturated fatty acids and *trans*-unsaturated fatty acids should represent ≤10% of each day's nutrition for the general population.[1,7,8] Saturated fats are the principal determinant of LDL-C levels, and *trans*-unsaturated fatty acids can also increase LDL-C, as well as reduce HDL-C levels. Foods high in saturated fat include full-fat dairy products, fatty meats, and tropical oils. In their place, polyunsaturated and monounsaturated fatty acids (eg, vegetable oils) are recommended. *Trans*-unsaturated fatty acids, which are found in the partially hydrogenated vegetable oils used for commercially prepared foods, have been more difficult to monitor than saturated fats, because food manufacturers were not required to list *trans*-fatty acid content. However, the Food and Drug Administration (FDA) recently issued new regulations that will require manufacturers of most conventional foods and some dietary supplements to include *trans*-fatty acids in their nutritional labeling by January 1, 2006.[9]

To a lesser extent than saturated and *trans*-unsaturated fats, dietary cholesterol is also associated with increased LDL-C levels. Since most foods high in saturated fat also contain cholesterol, reduced consumption of such products (eg, animal fats) limits cholesterol intake. The recommended intake of dietary cholesterol for the general population is <300 mg/d.

A high carbohydrate intake (>60% of total calories) should be avoided, because carbohydrates can result in elevated TG and reduced HDL-C levels. These undesirable effects can

Table 4-3: Nutrient Composition of the American Heart Association Diet

Nutrient	Recommended Intake
Dietary fats	≤30% of total calories
Saturated fats/ *trans*-unsaturated fatty acids*	≤10% of total calories
Cholesterol**	<300 mg/d
Carbohydrate***	≤60% of total calories

*For individuals with elevated LDL-C levels, diabetes/insulin resistance, or cardiovascular disease, the recommended daily intake is <7% of total calories. Currently, intake of *trans*-unsaturated fatty acids can be difficult to monitor because the content is not included in food labeling. As of January 1, 2006, however, new FDA regulations will require labeling of *trans*-fatty acid content.

**Lower intake (<200 mg/d) is recommended for individuals with elevated LDL-C levels, diabetes/insulin resistance, and/or cardiovascular disease.

***Including ≥25 g/d of grain products, with emphasis on whole grains and soluble fiber.

be lessened, however, when carbohydrates are derived from unprocessed whole foods (eg, fiber), rather than from monosaccharides (eg, fructose) and refined food products. Certain soluble fibers (eg, oat products, psyllium, pectin, guar gum) also reduce LDL-C, particularly in persons who are hypercholesterolemic. Because dietary fiber slows gastric emptying, it promotes a feeling of satiety that may help patients consume less food, thereby controlling their weight.

According to the AHA, there is insufficient evidence to know whether raising HDL-C levels and lowering TG concentrations can also decrease CHD risk. Consequently, these criteria are not included among the specific goals of dietary therapy. The guidelines point out, however, that a reciprocal metabolic relation exists between low HDL-C and elevated TG levels, and that both are associated with excess body weight, reduced physical activity, and intake of sugar and refined carbohydrates.

High-fiber dietary patterns have been associated with a reduced risk for CVD. Moreover, dietary fiber can help control calorie intake by promoting satiety, thereby contributing to weight loss. The AHA endorses a diet that includes five or more servings per day of fruits and vegetables, particularly those that are dark green, deep orange, or yellow; nuts and legumes; and six or more servings per day (≥25 g) of grain products, with emphasis on whole grains and soluble fiber. All of these foods, as well as low-fat dairy products, poultry, fish, and limited salt intake (no more than 6 g, or 2,400 mg, of sodium per day), have also been shown to help lower blood pressure. The AHA recommends that potassium, magnesium, and calcium, which are particularly important in preventing hypertension, come from food sources rather than mineral supplements. The one exception concerns women, who may need calcium supplementation to prevent or treat osteoporosis.

Medical Nutrition Therapy

As explained earlier, the AHA advocates medical nutrition therapy for persons at higher risk for CVD.

Four groups of people are identified as requiring more stringent nutritional and medical supervision (Table 4-2). The first group consists of persons with *elevated LDL-C levels* (ie, based on the ATP guidelines) *or preexisting CVD*. They should reduce their daily intake of saturated fat and cholesterol to <7% of total calories and <200 mg/d, respectively (Table 4-3). Appropriate monitoring is needed to determine if the diet is effective in decreasing LDL-C levels, to ensure that nutritional requirements are being met, and, in patients taking lipid-lowering medication, to decide if the dosage can be reduced. A fasting lipoprotein profile is the most effective means of evaluating lipid response to dietary and drug therapy. Patients whose total fat intake constitutes <15% of daily calories may have correspondingly increased their carbohydrate consumption; therefore, they should be monitored for the potential elevations in TG and reductions in HDL-C levels that may result.

Persons with diabetes and/or insulin resistance constitute the second group requiring medical nutrition therapy. Type 2, the most common form of diabetes, is associated with a twofold to fourfold increase in risk for CHD.[10] It is also associated with the metabolic syndrome, whose characteristics include central obesity, insulin resistance, dyslipidemia (ie, elevated TG; low HDL-C; small, dense LDL particles), hypertension, and a prothrombotic state. Even limited weight loss can lessen insulin resistance and its accompanying metabolic abnormalities, thereby modifying the risk for CVD. Patients in this group should also reduce daily intake of saturated fat to <7% of total calories and cholesterol consumption to <200 mg/d.

The relationship between poor glucose control and CVD has not been clearly defined. While hyperglycemia has been associated with an increase in cardiovascular risk,[11,12] pharmacologic lowering of blood glucose levels in persons with type 2 diabetes does not appear sufficient to reduce macrovascular complications.[13,14] The ongoing Action to Control Cardiovascular Risk in Diabetes (ACCORD) trial, scheduled for completion in September 2010, will evaluate whether lowering blood glucose reduces cardiovascular risk (see Chapter 6).[15]

Excess body fat (in particular, abdominal obesity) and physical inactivity promote the development of insulin resistance.[2] For persons with metabolic abnormalities associated with insulin resistance (eg, type 2 diabetes, risk factors for the metabolic syndrome), the AHA recommends a weight-reduction diet that replaces saturated fats with unsaturated fats, rather than with carbohydrates. Both unsaturated fats and carbohydrates can lower TC and LDL-C.[1] However, unsaturated fats offer additional cardioprotection by increasing HDL-C and decreasing TG compared with carbohydrates (see Adult Treatment Panel III Guidelines/Therapeutic Lifestyle Changes, this chapter).[16] In these patients, increased physical activity is also important for weight loss, raising HDL-C levels, and reducing VLDL-C levels.

Persons with congestive heart failure belong to the third group for which the AHA recommends medical nutrition therapy. Epidemiologic data indicate that hospitalization for congestive heart failure precedes two thirds of all deaths.[1] Nutritional factors can often affect the course of heart failure. Reducing sodium consumption may prevent exacerbation and enable the physician to lower the required dose of diuretic medication. Restricting water intake is advised, particularly in advanced cases. For patients who have right-sided heart failure related to obesity or sleep apnea, weight loss is a generally accepted therapeutic strategy.

The fourth group requiring medical nutrition therapy consists of *patients with kidney disease*. Almost one half of all kidney dialysis patients die of cardiovascular complications. Persons with renal dysfunction have a high incidence of diabetes, dyslipidemia (particularly hypertriglyceridemia), and hypertension. Renal disease patients may also have increased levels of serum lipoprotein(a) [Lp(a)] and homocysteine. An important dietary consideration in dialysis patients is to maintain muscle mass and limit hypoalbuminemia, which is associated with mortality. This can be done through the intake of protein-rich foods. However, saturated fat and cholesterol content must be restricted. In contrast, patients with progressive renal failure usually require a low-protein diet that is also low in salt but has an increased number of total calories.

Specific Nutrients

The AHA guidelines advocate further research into the effects of certain dietary factors on the development and expression of cardiovascular disease. Specifically mentioned are antioxidants, B vitamins and folic acid (including relationship to homocysteine), soy proteins and isoflavones, fiber, stanol/sterol esters, and ω-3 fatty acids derived primarily from fish oils.

In a more recent report, the AHA confirmed its position that scientific data do not justify the use of antioxidant vitamin supplementation to reduce cardiovascular risk.[17] The position of the AHA on ω-3 supplementation and on plant stanol/sterol ester-containing foods is discussed below.

Phytostanols and Phytosterols

Sterols are the alcohol form of steroids. C-28 and C-29 sterols occur naturally in plants and vegetable oils. They differ from cholesterol, which is a C-27 sterol, by the presence of an extra methyl or ethyl group on the cholesterol side chain. Stanols are hydrogenation compounds of sterols. Both stanols and sterols reduce plasma levels of TC and LDL-C by inhibiting absorption of dietary and biliary cholesterol from the intestinal tract, apparently by displacing cholesterol. Compared with cholesterol, plant sterols and stanols are relatively unabsorbable.[18,19] To receive a cholesterol-lowering benefit from stanols or sterols, a person must consume approximately 1 g/d. Since normal dietary intake of sterols provides just 200 to 400 mg/d (stanol intake is negligible), researchers are developing 'functional foods' that are enriched with plant stanols and sterols.[18]

Effects on lipids. β-Sitosterol is one of the most common phytosterols. Clinical trials in primary and secondary prevention have shown that margarine enriched with sitostanol ester, which is less absorbable than sitosterol, can lower serum levels of TC and LDL-C by 8% to 10% and by 14% to 15%, respectively, compared with a margarine lacking stanol enrichment.[20,21] When used in conjunction with simvastatin (Zocor®), the stanol-enriched margarine reduced serum TC and LDL-C levels by an additional 11% to 16%, respectively.[21]

Food and Drug Administration labeling. In September 2000, the FDA issued an interim final rule stating that stanol- and sterol-containing foods can qualify for labeling concerning a cardioprotective role. Eligible foods must contain at least 0.65 g of plant sterol esters or at least 1.7 g of plant stanol esters per serving and be low in saturated fat and dietary cholesterol, with no more than 13 g of total fat per serving and per 50 g (spreads and salad dressings may exceed the limit per 50 g if this is clearly indicated). In addition, the label must specify that the product should be consumed with food in two servings at different times of the day as part of a diet low in saturated fat and cholesterol.[22] Benecol® and Take Control®, two margarine spreads that meet these requirements, have been introduced into the US market. It has been reported that Benecol® and Take Control® have minimal amounts of *trans*-fatty acids (eg, <0.5 g per 8-g serving of Benecol®[23]). More recently, plant sterol-fortified orange juice has also become available.

The FDA is considering comments and additional scientific evidence before issuing its final rule.[24,25]

American Heart Association position. According to an AHA advisory report, plant stanol/sterol ester-containing foods should be reserved for use in adults requiring reduction of TC and LDL-C levels because of hypercholesterolemia or the need for secondary prevention of atherosclerotic events. In children with hypercholesterolemia, these foods can be considered as an adjunct to diet, but fat-soluble vitamin status must be monitored. One concern about widespread availability of stanol/sterol ester-fortified foods is that they may be shared with family members at meals. Long-term safety and efficacy studies are needed in both children and adults, including the general population, pregnant women, and normocholesterolemic individuals with other CHD risk factors. Issues include a possible decrease in levels of fat-soluble nutrients (eg, β-carotene, α-tocopherol, lycopene) and the biological significance of such effects.[24] Although plasma levels of plant sterols/stanols are not elevated or only minimally elevated after ingestion of these foods, it is not clear how individuals with abnormally high plant sterol absorption (eg, those who are homozygous for sitosterolemia) may respond. In addition, before fortification of the food supply can be considered as a population-wide means of reducing coronary risk, cost-effectiveness data are needed.[24]

Fish Oil

Fish oil is another promising dietary intervention. A growing body of evidence indicates that the ω-3 fatty acids contained in fish oil—eicosapentaenoic acid (EPA), docosahexaenoic acid (DHA) and, in some instances, α-linolenic acid—are variously associated with lowering plasma TG levels, decreasing the risk for arrhythmia, reducing the incidence of sudden cardiac death and myocardial infarction (MI), slowing the progression of coronary artery disease, and inhibiting platelet aggregation.[27-36] In addition, autopsy studies performed predominantly in Greenland Eskimos and Alaskan natives found a correlation between a diminished prevalence of both coronary and peripheral atherosclerosis and increased levels of dietary EPA and DHA.[37]

Mechanisms of action. There are several mechanisms by which ω-3 fatty acids may alter the development or progression of CHD. While some mechanisms appear to involve lipid metabolism, others may be related to smooth muscle cell proliferation, platelet aggregation, and vasoconstriction.

The lipid effects of ω-3 fatty acids may include a decrease in the secretion of apolipoprotein (apo) B, leading to a decline in the rates of secretion of TG and VLDL. It has also been suggested that ω-3 fatty acids may increase lipoprotein degradation. One experimental model in rats found that dietary consumption of EPA led to profound changes in TG synthesis that did not occur when EPA was added in vitro to the hepatocytes of rats fed a low-fat diet.[38-43] The inhibitory effect of ω-3 fatty acids on lipid metabolism in vivo may be partly related to induction of peroxisome proliferator-activated receptor α, a nuclear transcription factor for which fatty acids are a ligand.[38,44]

The nonlipid atheroprotective actions of EPA and DHA may involve alterations in the effects of thromboxane A_2, which include vasoconstriction and promotion of platelet aggregation. Both EPA and DHA have been shown to prevent smooth muscle cell proliferation induced by thromboxane A_2,[45] and DHA appears to be a competitive inhibitor of prostaglandin synthetase-mediated conversion of arachidonic acid to thromboxane.[46] EPA and DHA may also limit platelet aggregation via a mechanism that could be related to a decrease in the effectiveness of thromboxane A_2.[47-50] In addition, researchers have hypothesized that ω-3 fatty acids can enhance the endothelin-dependent relaxation of coronary arteries.[51]

Effects on lipids. The effect of fish oils on circulating lipids varies and, at least to some extent, depends on the pretreatment lipid phenotype. Fish oil supplementation can decrease serum TG levels by 20% to 30% and VLDL by 30% to 40% compared both with placebo and with baseline.[52] Fish oil supplements and dietary fish oil have differing effects on HDL-C but generally produce a modest increase.[53-56] The variation in HDL-C response may be related to the apo E genotype.[55]

In patients who are hypertriglyceridemic, fish oil therapy has been shown to decrease plasma concentrations of total TG, as well as the TG content of VLDL and LDL particles. While plasma levels of TC and VLDL-C also decreased, concentrations of LDL-C and HDL-C increased.[57] The elevation in LDL-C levels, which may indicate the presence of an underlying genetic defect,[58] correlates with the presence of small, dense apo B-rich LDL particles that become cholesterol-enriched and more buoyant with fish oil supplementation, but retain their atherogenic size. In patients with hypertriglyceridemia, there also appears to be decreased LDL binding to fibroblast receptors, which could help explain the increase in LDL-C concentrations,[57] as well as an enhanced susceptibility of LDL to oxidation in vitro. The latter effect would not necessarily promote lesion development in vivo, however, because it may be attenuated by the presence of antioxidative mechanisms.[59]

Clinical efficacy. The GISSI-Prevenzione trial[35] compared supplements of fish oil (Omacor®), vitamin E, or both vs no supplementation in 11,324 patients with recent (≤3 months) MI. Results indicate that only fish oil led to a significant decrease (3.4%) in TG concentrations compared with controls. According to a four-way analysis, ω-3 fatty acids decreased the risk for death, nonfatal MI, and stroke (combined primary end point) by 15% (P=0.023) and for cardiovascular death, nonfatal MI, and nonfatal stroke by 20% (P=0.008). Within the primary end point, the risk for total deaths decreased by 20%, cardiovascular deaths by 30%, and sudden deaths by 45% with fish oil supplementation. There was no difference between vitamin E and control in the combined primary end point, although the decreases in cardiovascular death and sudden death (20% and 35%, respectively) were similar to those with fish oil.

The outcome of the GISSI study is similar to that of the Diet and Reinfarction Trial (DART),[60] in which 2,033 men with prior MI received dietary advice about fatty fish consumption, fiber consumption, or reduction in fat intake. Each group was compared with men who received no dietary advice. The subjects receiving advice on fat consumption had lower serum cholesterol levels than those in the other groups. Although none of the groups experienced significant decreases in ischemic heart disease compared with controls or unity, fatty fish consumption reduced total mortality by approximately 29% during the first 2 years after an MI.

Side effects and drug interactions. Gastrointestinal complaints, such as moderate diarrhea, are common with fish oil supplements. The increased bleeding time that results from inhibition of platelet aggregation has potential dangers, especially for patients who are concomitantly taking anticoagulants, including aspirin. Fish oil supplements may contain potentially toxic amounts of the lipid-soluble vitamins A and D, so patients should be monitored for potential adverse effects, such as dermatitis or hypercalcemia.

Food and Drug Administration labeling. A number of European countries have approved the use of Omacor®, a form of esterified ω-3 fatty acids found to reduce cardiovascular risk in post-MI patients.[35] In the United States, fish oil supplements are available as over-the-counter products. A decision by the FDA in 2004 permits the use of a 'qualified' health claim on foods containing EPA and DHA (a similar qualified claim for dietary supplements was approved in 2000).[61] Qualified health claims are based on data that appear promising but inconclusive.[62]

The following wording has been approved: "Supportive but not conclusive research shows that consumption of EPA and DHA ω-3 fatty acids may reduce the risk of coronary heart disease. One serving of [name of food] provides [x] grams of EPA and DHA ω-3 fatty acids. (See nutrition information for total fat, saturated fat, and cholesterol content.)"[61] The FDA recommends that consumption of ω-3 fatty acids not exceed 3 g/d, with no more than 2 g/d from a dietary supplement (a minimum recommended amount is not stipulated).[61] To prevent food manufacturers from simply adding EPA and DHA to otherwise unhealthy products, labeled foods (with the exception of fish and dietary supplements) must also have <20 mg of cholesterol and <1 g of saturated fat per serving, with <15% of total calories from saturated fat.[62]

Some public advocacy groups object to qualified health claims, believing that they are confusing for consumers.[62]

American Heart Association position. In a scientific statement on fish consumption, ω-3 fatty acids, and CVD, the AHA suggests that patients with coronary artery disease may, in consultation with their physicians, consider ω-3 fatty acid supplementation for the reduction of coronary risk; supplementation may also be appropriate for the medical management of hypertriglyceridemia.[63]

Adult Treatment Panel III Guidelines

To meet the primary goal of lowering LDL-C levels, as well as to modify other CHD risk factors, the ATP III guidelines present nonpharmacologic (lifestyle) and pharmacologic strategies.[2,3] This chapter reviews the nonpharmacologic recommendations.

Therapeutic Lifestyle Change: Description

Therapeutic lifestyle change (TLC) consists of the following: reduction in saturated fat and cholesterol intake, increase in consumption of plant stanols/sterols and soluble fiber, increase in physical activity, and weight loss. Lifestyle changes are the basis of CHD prevention. Although persons at higher risk require intensified prevention, including drug therapy, everyone with elevated LDL-C or lifestyle-related risk factors should adhere to appropriate lifestyle changes. For the metabolic syndrome, which increases the risk for CHD at any LDL-C level, intensified weight loss and physical activity are considered first-line therapy.

The TLC diet is based on the Dietary Guidelines for Americans 2000, with some modifications.[64] Fat allowance is 25% to 35% of total calories, with the proviso that intake of saturated fats and *trans*-fatty acids be kept low (See American Heart Association/Dietary Recommendations). This is warranted because higher consumption of total fat, primarily unsaturated fat, can help reduce TG levels and increase HDL-C levels in patients with the metabolic syndrome. An exception is made for patients with TG levels ≥500 mg/dL, whose fat intake should be reduced to ≤15% of total calories. The nutrient composition of the TLC diet is shown in Table 4-4. Moderate physical activity is also recommended. Table 4-5 presents the TLC approach to reducing elevated TG levels, together with the recommended goals of therapy. The goal of TLC is twofold: to lower LDL-C levels and to reduce risk via other mechanisms.[3] Therefore, high- or moderately high-risk patients with lifestyle-related risk factors or the metabolic syndrome are candidates for TLC regardless of LDL-C level.[3]

Figure 4-1 illustrates the TLC model. In the initial visit, the physician should emphasize the importance of reduc-

Table 4-4: Nutrient Composition of the TLC Diet

Nutrient	Recommended Intake
Saturated fat*	<7% of total calories
Polyunsaturated fat	Up to 10% of total calories
Monounsaturated fat	Up to 20% of total calories
Total fat**	25% to 35% of total calories
Carbohydrate***	50% to 60% of total calories
Fiber	20 to 30 g/d
Protein	15% of total calories (approximately)
Cholesterol	<200 mg/d
Total calories	Balance energy intake and expenditure to maintain desirable body weight/prevent weight gain. Daily energy expenditure should include at least moderate physical activity (contributing approximately 200 kcal/d).

* Intake of *trans*-fatty acids, another cholesterol-raising fat, should also be low. Currently, this can be difficult to monitor because it is not included in food labeling. As of January 1, 2006, however, new FDA regulations will require labeling of *trans*-fatty acid content.

** For patients with elevated TG levels (≥500 mg/dL), fat intake should be ≤15% of total calories.

*** Derived predominantly from foods rich in complex carbohydrates, including grains (especially whole grains), fruits, and vegetables.

Adapted from Executive Summary of the Third Report of the National Cholesterol Education Program (NCEP) Expert Panel on Detection, Evaluation, and Treatment of High Blood Cholesterol in Adults (Adult Treatment Panel III).[2]

TLC = therapeutic lifestyle change; TG = triglycerides

Table 4-5: TG Goals and TLC Recommendations

Classification (mg/dL)	Nondrug Therapy	TG Goal (mg/dL)
Normal (<150)	Begin/maintain TLC (to prevent elevated TG)	<150
Borderline-high (150-199)	Begin/maintain TLC, including: • weight reduction* • physical activity* • smoking cessation • decreased carbohydrate intake (≤60% of daily calories) • control of alcohol intake Diagnose/treat underlying conditions that cause secondary TG elevations, such as: • metabolic syndrome • diabetes • chronic renal failure • nephrotic syndrome	<150
High (200-499)	All of the above	<150
Very high (≥500)	All of the above plus: • reduced fat intake (≤15% of daily calories)	<150

*First-line therapy for the metabolic syndrome. Elevated TG levels are usually associated with the metabolic syndrome.
TG = triglycerides; TLC = therapeutic lifestyle change

ing saturated fat and cholesterol intake. If the patient's LDL-C goal has not been achieved by visit 2, increased consumption of plant stanols/sterols and soluble fiber may be advised. After the LDL-C level has been reduced as much as possible with dietary therapy, the metabolic syndrome becomes the focus of attention, if not already addressed. Since most patients with this syndrome are overweight or obese and sedentary, intensified weight management and physical activity are the primary interventions. Both can improve the lipid and nonlipid abnormalities associated with the metabolic syndrome and may also help lower LDL-C levels. The first three visits are 6 weeks apart. Following visit 3, the patient should be seen every 4 to 6 months so that compliance can be monitored. At each step, the physician may decide to refer the patient to a qualified dietitian or nutritionist for medical nutrition therapy.

Therapeutic Lifestyle Change: LDL-C Cut Points and Goals

Table 4-6 presents LDL-C cut points for initiating TLC and the goals to be achieved. In high-risk patients, TLC is recommended whenever the LDL-C level is ≥100 mg/dL. According to the revised ATP III guidelines, however, high-risk patients with lifestyle-related risk factors or the metabolic syndrome are candidates for TLC regardless of LDL-C level. The guidelines also recommend simultaneously initiating TLC and drug therapy in all high-risk patients with an LDL-C level of 100 to 129 mg/dL or higher (compared with ≥130 mg/dL in the 2001 guidelines).[3]

Patients with ≥2 major risk factors are divided according to their 10-year likelihood of developing CHD based on the Framingham scoring system: 10% to 20% (moderately high risk) or <10% (moderate risk). In those at moderately high risk, TLC should be initiated whenever the LDL-C level is ≥130 mg/dL. If the LDL-C level remains at that level after a trial of TLC, the physician can consider initiating an LDL-lowering drug in order to achieve a goal of <130 mg/dL. Based

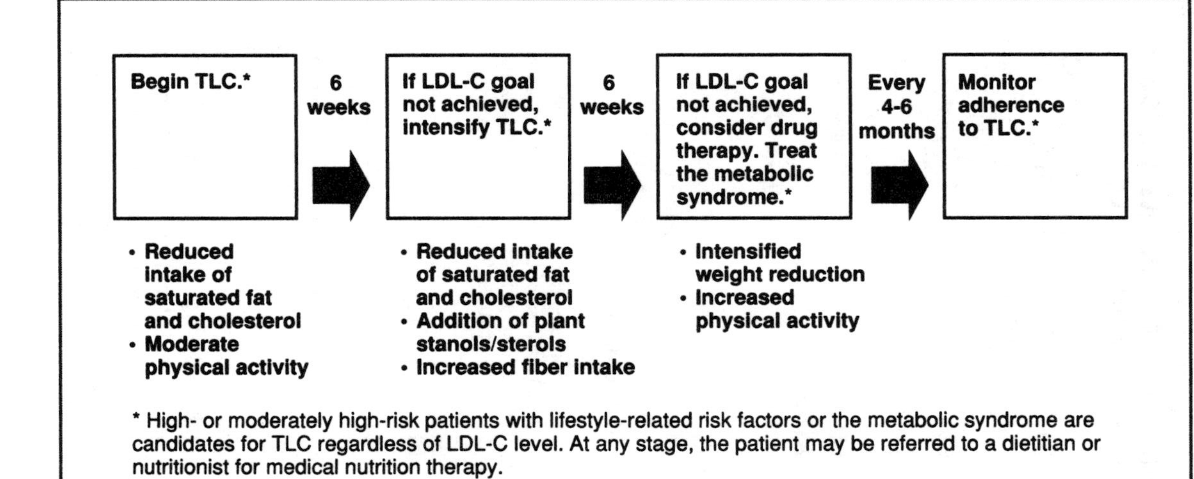

Figure 4-1: Model of steps in TLC. *JAMA* 2001;285:2486-2497. *Circulation* 2004;110:227-239.

Table 4-6: LDL-C Cut Points and Goals for Initiating TLC

Risk Category	LDL-C (mg/dL) Cut Point (Goal)
1. High risk*	
CHD or CHD risk equivalent (>20% risk)	≥100 (<100; optional goal <70)
2. Moderately high risk*	
Multiple (≥2) risk factors (10% to 20% risk)	≥130 (<130; optional goal <100)
3. Moderate risk	
Multiple (≥2) risk factors (<10% risk)	≥130 (<130)
4. Lower risk	
0-1 risk factor Most patients with 0-1 risk factor have a 10-year risk <10%	≥160 (<160)

* In patients with lifestyle-related risk factors or the metabolic syndrome, initiate TLC regardless of LDL-C level. Simultaneous initiation of TLC and drug therapy is recommended in high-risk patients with LDL-C levels of 100-129 mg/dL or higher.
CHD = coronary heart disease
LDL-C = low-density lipoprotein cholesterol
TLC = therapeutic lifestyle changes

on clinical trial evidence, there is also the option of initiating LDL-lowering drug therapy if the LDL-C level is 100 to 129 mg/dL on TLC, with a goal of <100 mg/dL. As in the case of high-risk patients, the revised guidelines recommend TLC for any moderately high-risk patient with lifestyle-related risk factors or the metabolic syndrome, regardless of LDL-C level.[3] In patients with ≥2 risk factors and a <10% chance of developing CHD over 10 years, TLC remains the primary intervention as long as the LDL-C level remains

<160 mg/dL (≥160 mg/dL is the cut point for initiating drug therapy in this group).

For most patients with 0 to 1 risk factor, the 10-year probability of developing CHD is <10%. In these patients, TLC is the principal risk-reduction strategy. After 3 months, if LDL-C levels are <160 mg/dL, then TLC is continued. If LDL-C levels are ≥160 mg/dL, or if certain other circumstances are present, drug therapy may be considered cost-effective to slow the long-term development of coronary atherosclerosis. Circumstances that may warrant drug therapy in addition to TLC include a severe single risk factor (eg, heavy cigarette smoking, poorly controlled hypertension), multiple life-habit or emerging risk factors, or LDL-C levels ≥190 mg/dL.

Dietary Therapy in Clinical Practice

Efficacy

Epidemiologic evidence. Epidemiologic data clearly indicate that there is an association between diet and CHD. The Seven Countries Study, which began in 1958, surveyed 12,763 men in seven countries (the United States, Finland, The Netherlands, Italy, former Yugoslavia, Greece, and Japan) over a 25-year period to investigate the relationship between dietary habits and the incidence of sudden coronary death or fatal MI.[65] Based on a comparison of 18 different simple food groups and three combined food groups (vegetable, animal, sweets), researchers found that all animal foods (including dairy and meat) and sweets were directly correlated with CHD death rates, whereas vegetable foods (except potatoes and fruit), alcohol, and fish were inversely related to CHD mortality. The following high-consumption dietary patterns were identified: dairy products in the northern European diet; meat in the United States; vegetables, cereals, olive oil, fish, and wine in the Mediterranean region; and rice, soy products, and fish in Japan. In a subset of 11,579 men aged 40 to 59 years who were identified as 'healthy' at baseline, 2,288 died within 15 years, with 80% of the deaths caused by CHD.[66] Among these subjects, there was a positive association between death rates and saturated fatty acid consumption, a negative association between death rates and intake of monounsaturated fatty acids, and no relationship between death rates and consumption of polyunsaturated fatty acids, proteins, carbohydrates, and alcohol. Furthermore, CHD death rates were low in subjects whose primary dietary fat was olive oil.

Using a food-frequency questionnaire, investigators from the Harvard School of Public Health conducted a prospective cohort study assessing the relation between diet and the incidence of nonfatal MI or fatal CHD in 84,251 women and 42,148 men from the Nurses' Health Study and the Health Professionals' Follow-Up Study.[67] They found that the highest quintile of fruit and vegetable intake was associated with a 20% reduction in risk for CHD compared with the lowest quintile. Furthermore, there was a 4% lower risk for CHD with each 1 serving/day increase in fruits or vegetables. Green leafy vegetables and fruits or vegetables rich in vitamin C appeared to offer the most cardioprotection. Data analyses from the Framingham Heart Study found that fruit and vegetable intake was also associated with a protective effect in ischemic stroke.[68]

Interventional trial evidence: clinical end points. Based on results of the Seven Countries Study, the 'Mediterranean diet' has been publicized as a means of reducing cardiovascular risk. The Lyon Diet Heart Study, a randomized, single-blind trial, compared a Mediterranean-type diet with a prudent diet in 605 subjects recovering from an MI. After a mean of 27 months, the Mediterranean diet was associated with a decrease of approximately 70% in risk for a primary end-point event (cardiovascular death/nonfatal MI).[69] Total mortality also decreased by 70%. The protective effects of the Mediterranean diet were related to changes in plasma fatty acids (ie, increases in ω-3 fatty acids and oleic acid, a decrease in linolenic acid). Plasma concentrations of antioxidant vitamins C and E increased significantly (by week 52) in the experimental group, but serum lipids, body mass index (BMI), and blood pressure remained similar with both interventions. A final follow-up confirms the results of this trial (see American Heart Association Position, below).

Interventional trial evidence: risk factors. Although there is a paucity of interventional trial data showing that coronary disease can be prevented by diet alone, evidence suggests that nutrition can influence the presence of cardiovascular risk factors.

In a study comparing diet with exercise, 157 healthy men aged 35 to 60 years were randomized to diet, exercise, diet + exercise, or no active intervention.[70] As estimated by the Framingham scoring system, 10-year risk in the diet, exercise, and diet + exercise groups was reduced by 13%, 12%, and 14%, respectively. These results were accompanied by decreases in BMI, waist circumference, TC, LDL-C, and VLDL-C. High-density lipoprotein cholesterol and TG levels were not affected. The investigators concluded that diet and exercise can modify several important cardiovascular risk factors and that both modalities were about equally effective in reducing the 10-year estimated risk for CVD. The subject of exercise is discussed later in this chapter.

A meta-analysis of 37 randomized dietary intervention studies published between 1981 and 1997 found that the NCEP Step I and Step II guidelines were effective in reducing blood cholesterol concentrations.[71] A total of 9,276 subjects and 2,310 controls participated in the trials. The Step I and Step II diets led to mean decreases in plasma levels of TC (10% and 13%, respectively), LDL-C (12% and 16%), and TG (8% and 8%).[70] For TC and LDL-C, these effects were additive, with incremental decreases of 2.3 mg/dL and 1.6 mg/dL, respectively, for every 1% reduction in calories from dietary fat. Furthermore, subjects who exercised had greater reductions in TC, LDL-C, and TG levels. Although HDL-C levels decreased with both the Step I and Step II diets, increases occurred in subjects who lost weight.[70]

In a small trial of 32 men and women (8 with LDL-C <160 mg/dL and 24 with LDL-C ≥ 160 mg/dL), a baseline

'American' diet (ie, 35% of calories from fat, 14% from saturated fat, 147 mg cholesterol/1,000 kcal) for 6 weeks was followed by a Step II diet.[72] In both groups, the Step II diet was associated with significant reductions in TC (20% and 16%, respectively) and LDL-C (21% and 18%). Levels of HDL-C also decreased.

More recently, the effects of three different food regimens on plasma lipids were compared in the Dietary Approaches to Stop Hypertension (DASH) trial, a multicenter outpatient study.[73] A total of 436 participants (60% African American) with baseline TC ≤259 mg/dL were randomized to 8 weeks of a control diet, a diet increased in fruits and vegetables, or the DASH diet (ie, increased fruits, vegetables, and low-fat dairy products and decreased saturated fat, total fat, and cholesterol). Compared with the control diet, the DASH diet was associated with lower levels of TC (14 mg/dL), LDL-C (11 mg/dL), and HDL-C (4 mg/dL). There were no significant effects on TG concentrations. With the fruit and vegetable diet, there were no significant reductions in TC, LDL-C, or HDL-C, but a significant decrease was found in TG concentrations.

Interventional trial evidence: metabolic syndrome. Another study indicates that risk factors for the metabolic syndrome (ie, insulin resistance, elevated TG levels) are responsive to dietary intervention. A group of 24 'healthy' obese women (BMI >30.0 kg/m^2) was divided at baseline into insulin-resistant and insulin-sensitive subgroups based on steady-state plasma glucose concentrations.[74] Baseline TG levels and the ratio of TC to HDL-C were higher in the insulin-resistant women. After 4 months of a calorie-restricted diet plus the antiobesity drug sibutramine (Meridia®), there was significant weight loss in both groups. However, significant reductions in steady-state plasma glucose concentrations, day-long plasma glucose and insulin concentrations, and fasting TG levels were observed in the insulin-resistant subjects only. These results suggest that CHD risk factors in obese women are related to insulin resistance and that weight loss reduces CHD risk in obese women who are insulin resistant. The latter conclusion is consistent with the ATP III recommendations that weight reduction (and intensified physical activity) are first-line therapy for the metabolic syndrome.

Nonlipid risk factors. In addition to plasma glucose and insulin concentrations, other nonlipid factors may respond to nutritional intervention. One promising area is endothelial dysfunction, which is found both in patients with CVD (eg, CHD, peripheral arterial disease, chronic heart failure) and in those without established CHD but with hypercholesterolemia, major CHD risk factors (eg, smoking, hypertension), hyperhomocysteinemia (an emerging risk factor), or diabetes (a CHD risk equivalent). Endothelial dysfunction can be clinically assessed by measuring plasma concentrations of soluble endothelial adhesion molecules or determining endothelium-dependent vasodilatation.[75] In an extensive review of laboratory, clinical experimental, prospective cohort, and epidemiologic studies, Brown and Hu suggest that ω-3 fatty acids, antioxidant vitamins, folic acid, and L-arginine may have beneficial effects on endothelial function.[75] The possible mechanisms include inhibiting monocyte adhesion and platelet activation, improving vasodilatation by increasing nitric oxide production, and blocking lipid oxidation.

The ATP III guidelines also include proinflammatory factors among the emerging risk factors for CVD and the metabolic syndrome. C-reactive protein (CRP), an inflammatory response protein, was the strongest univariate predictor of cardiovascular events in a nested case control study comparing the predictive value of 12 markers of inflammation in 28,263 women.[76] Because CRP is also associated with BMI in obese women, a study was conducted to determine if weight loss can reduce CRP concentrations.[77] After 12 weeks of a low-fat, energy-restricted diet, CRP levels were reduced by 26% in 83 apparently healthy obese women. Further study is needed, however, to determine if this effect is associated with reductions in atherosclerosis progression.

Atherosclerosis is a complex process whose rate of development at any given lipid level is influenced by many nonlipid factors. These nonlipid mechanisms may also be involved in the pathogenesis of atherosclerosis at the cellular-molecular level.[78-80] In addition to improving our understanding of the effects of diet, weight loss, and exercise on nonlipid risk factors, we must encourage further research into the crucial relationship between genetics and individual response to dietary change.[81-83]

American Heart Association Position

In a science advisory discussing the final report of the Lyon Diet Heart Study, the AHA advocates further nutrition-based research. After 46 months of follow-up, this study of a Mediterranean-type diet (in the context of a Step I regimen) vs a prudent diet found that free-living subjects with a history of MI experienced significant reductions of 65% in cardiac deaths and 72% in the composite primary end point of MI plus cardiovascular death.[84] These results are consistent with the 27-month interim analysis (see Efficacy/Interventional trial evidence: clinical end points, above).[84,85] There was also a 56% reduction in total mortality. If the Mediterranean-type diet had been prescribed as part of a Step II program, the outcomes might have been even more impressive.[85]

Despite several methodologic limitations, these findings illustrate the potential importance of a diet rich in fruits, vegetables, breads and cereals, and fish, as well as α-linolenic acid.[85] Because TC, LDL-C, and HDL-C levels at the final visit were similar in the experimental and the control groups, it is reasonable to suggest that ω-3 fatty acids have multiple cardioprotective mechanisms. These may include antiarrhythmic effects, anti-inflammatory/antithrombotic properties, stimulation of endothelial-derived nitric oxide, and inhibition of cytokine synthesis. The results of the Lyon Diet Heart Study, if corroborated, are potentially of enormous significance for cardiovascular risk reduction in the general population. Therefore, it is important for the AHA and its partners to fund future research to determine if these results can be replicated and to identify the specific effects of each dietary component.[85]

Patient Compliance

Patient compliance is a significant consideration in dietary and other lifestyle modifications, and some studies have investigated methods of improving adherence to the recommended regimen. The Cholesterol Lowering Intervention Program (CLIP) was designed by the National Heart, Lung, and Blood Institute to test the feasibility of implementing Adult Treatment Panel I (ATP I) guidelines in physicians' offices.[86] Despite the fact that these guidelines have been superseded by more recent recommendations, the CLIP findings are still informative. In CLIP, the efficacy of three protocols was assessed in 23 physician practices treating a total of 480 hypercholesterolemic patients: usual care, an office-assisted model, and a nutrition-center model. For the office-assisted model, physicians and their office personnel attended a 1-day seminar that focused on dietary therapy for treating hypercholesterolemia, compliance techniques, and counseling skills. In the nutrition-center model, physicians identified patients with high serum cholesterol levels and referred them to a nutritionist for treatment. Compared with baseline, mean cholesterol response was significant across the three models in patients not taking lipid-lowering medication: usual care, decrease of 5.4 mg/dL ($P<0.01$); office-assisted model, decrease of 12 mg/dL ($P<0.05$); and nutrition-center model, decrease of 20.9 mg/dL ($P<0.05$). The change in the usual-care patients was significantly smaller than in the two enhanced-intervention models. Furthermore, the change in the nutrition-center model was significantly greater than that observed in the office-assisted model ($P<0.004$).

The mean percentage change in cholesterol in any model was less than 10%, indicating that even with physician support, dietary therapy achieves only modest cholesterol reductions. However, these results demonstrate that long-term monitored lipid-lowering programs can improve compliance and lead to serum cholesterol reductions greater than those achieved without professional support.

The beneficial effects of medical nutrition therapy were reported in a 3-month prospective randomized study involving 52 men and 52 women at risk for CVD.[87] The study compared dietary counseling by a physician or nurse with physician or nurse counseling plus medical nutrition therapy provided by a registered dietitian. While both groups experienced statistically significant reductions in serum lipid levels, the group with the services of a registered dietitian had significantly greater improvement in BMI, increased knowledge and understanding of nutrition, and lower consumption of fat, carbohydrates, and cholesterol. It is suggested that improved knowledge of nutrition and decreased intake of atherogenic foods may significantly promote long-term adherence to a cardioprotective diet.

According to a cross-sectional analysis of estimated nutrient intake for men (n = 1,145) and women (n = 1,375) in the Framingham Offspring-Spouse Study (1991 to 1995), adherence to nutritional guidelines for preventing disease and promoting health is improving, but more progress is needed.[88] Based on 3-day food records, results indicate that ≥70% of the sample met the recommendations for intake of polyunsaturated and monounsaturated fat, cholesterol, alcohol, vitamins C and B_{12}, and folic acid. However, ≤50% adhered to the guidelines for intake of carbohydrates, total and saturated fat, β-carotene, and vitamins A, E, and B_6. There were also important gender differences in nutrient intake, which suggests a need for gender-specific dietary messages and behavioral interventions to improve patient compliance.[88]

Physician Compliance

The issue of patient compliance is inextricably linked to that of physician compliance (see Chapter 3). Since 1988, studies have reported a great variation in the implementation of preventive cardiovascular services by physicians, with much need for improvement.[89-107] In the area of nutritional therapy, there are many reasons why physicians may not provide dietary counseling. These are summarized in Table 4-7.

The Coronary Artery Disease Reversal (CADRE) program has shown that educating internal medicine residents in dietary counseling increased the percentage who felt confident to provide this service from 26% to between 67% and 78%.[100] A prompting intervention (ie, a fingerstick blood cholesterol test before the patient's visit) approximately doubled the frequency of dietary counseling and increased the likelihood that patients would try to make the dietary changes. Despite these interventions, however, cholesterol levels decreased only marginally.

Although progress has been made in lowering cardiovascular risk,[103] much remains to be done. Among the diet-related issues that must be addressed are the increases in obesity and type 2 diabetes in the United States.[107] The ATP III guidelines recommend several strategies that can foster physician and patient compliance with risk-reduction strategies (Appendix B).

Exercise

The AHA Task Force on Risk Reduction has concluded that a substantial body of evidence points to the benefits of physical activity in both primary and secondary prevention.[108]

Physical exercise has been associated with a lower risk for CHD.[109] Recent studies demonstrate that for cardiac patients, exercise improves lipid profile. It also leads to decreases in BMI and percentage of body fat and to an increase in exercise capacity. In addition, exercise can induce psychological improvements (eg, decreased anxiety, hostility, and depression). These conditions have been associated with the progression of CHD.[110,111] The benefits of physical activity for all populations appear to be established, yet questions linger regarding the effect of intensity, the extent to which exercise affects blood lipids and apolipoproteins, and the precise relation between exercise and diet.

Physicians must understand that exercise and diet are paired interventions. For example, the combination of weight loss and diet has been shown to offset the decrease in HDL-C

Table 4-7: Physician Implementation of Dietary Guidelines: Factors That Can Hinder Compliance

Clinical

- Lack of information about patient risk (eg, lipid levels not measured)
- Misperception about patient risk (eg, cardiovascular risk in women)

Educational

- Insufficient training in:
  - nutrition
  - behavior modification
  - communication skills
  - ATP guidelines

Individual

- Lack of confidence in preparation or skill to help patients modify behavior
- Doubt about patient interest
- Concern about job satisfaction (lack of interest in counseling)
- Poor dietary habits, obesity, or other factors that hamper ability to serve as role model

Institutional

- Insufficient time
- Absence of systems and targets for improvement
- Lack of patient education materials and other tools
- Inadequate or no reimbursement for services
- Unclear role definition (eg, does responsibility lie with primary care physician or specialist?)

ATP = Adult Treatment Panel

Table 4-8: Smoking Intervention Guidelines

- Every person who smokes should be counseled on smoking each time he or she visits the physician's office. Maintaining smoking cessation should be frequently discussed with patients who have quit.
- Every patient should be asked about tobacco use; smoking status should be recorded and updated regularly.
- Cessation interventions as brief as 3 minutes are effective, with more intensive intervention being more effective.
- Clinicians should receive training in patient-centered counseling methods.
- Office systems should be established that facilitate smoking cessation intervention.
- Links with other personnel and organizations should be established to provide smoking cessation intervention (nurses, smoking cessation specialists, multiple risk factor intervention programs, community resources).

From Ockene IS, Miller NH: Cigarette smoking, cardiovascular disease, and stroke. A statement for healthcare professionals from the American Heart Association Task Force on Risk Reduction. *Circulation* 1997;96:3243-3247.

seen with the Step I diet,[112] and exercise without caloric restriction in hypercholesterolemic men appears to have a negligible effect on lipids, regardless of its intensity.[113]

Ideally, an exercise regimen should begin during youth and continue throughout life. The AHA recommends 30 minutes or more of moderate-intensity physical activity (eg, brisk walking) on most, and preferably all, days of the week.[108] Particularly in secondary prevention, exercise programs must be tailored to the patient's functional capacity. For patients who may be at high risk for CHD, such as those with hypertension or elevated cholesterol levels, periodic testing for functional capacity is recommended. In 2000, the AHA published its recommendations in a science advisory on resistance exercise in individuals with and without CVD (Table 4-1).[114]

Smoking and Alcohol Use

The AHA unequivocally endorses efforts to stop smoking. As many as 30% of all CHD deaths in the United States may be attributed to cigarette smoking, and the risk is dose-related.[115] The AHA has issued many position statements regarding the risks associated with tobacco use and the clinician's role in facilitating smoking cessation (Table 4-8). Patients who smoke should be encouraged to quit, and a number of programs and products are available to assist in smoking cessation. Generally, long-term quitters may gain about 5 to 6 kg (11 to 13 lb) in weight, although physical exercise may lessen the amount of weight gain.[116] Care must be taken to avoid an excessive increase in weight.[116]

Although several large observational studies have found a relationship between high blood pressure and consumption of three or more alcoholic beverages per day, data from some population studies indicate an association between moderate alcohol intake and a reduced number of cardiovascular events. Nevertheless, the adverse effects of alcohol outweigh its possible benefits, and the AHA does not recommend alcohol as a cardioprotective substance.[117]

References

1. Krauss RM, Eckel RH, Howard B, et al: AHA Dietary Guidelines. Revision 2000: A statement for healthcare professionals from the Nutrition Committee of the American Heart Association. *Circulation* 2000;102:2296-2311.

2. Expert Panel on Detection, Evaluation, and Treatment of High Blood Cholesterol in Adults: Executive Summary of the Third Re-

port of the National Cholesterol Education Program (NCEP) Expert Panel on Detection, Evaluation, and Treatment of High Blood Cholesterol in Adults (Adult Treatment Panel III). *JAMA* 2001; 285:2486-2497.

3. Grundy SM, Cleeman JI, Merz CN, et al: Implications of recent clinical trials for the National Cholesterol Education Program Adult Treatment Panel III guidelines. *Circulation* 2004;110:227-239.

4. Obarzanek E, Hunsberger SA, Van Horn L, et al: Safety of a fat-reduced diet: the Dietary Intervention Study in Children (DISC). *Pediatrics* 1997;100:51-59.

5. Jacobson MS: Heart healthy diets for all children: no longer controversial. *J Pediatr* 1998;133:1-2.

6. Tershakovec AM, Jawad AF, Stallings VA, et al: Growth of hypercholesterolemic children completing physician-initiated low-fat dietary intervention. *J Pediatr* 1998;133:28-34.

7. Kris-Etherton P, Daniels SR, Eckel RH, et al: Summary of the Scientific Conference on Dietary Fatty Acids and Cardiovascular Health. Conference summary from the Nutrition Committee of the American Heart Association. *Circulation* 2001;103:1034-1039.

8. Wylie-Rosett J: Fat substitutes and health. An Advisory from the Nutrition Committee of the American Heart Association. *Circulation* 2002;105:2800-2804.

9. US Department of Health and Human Services: HHS to require food labels to include trans fat contents. Improved labels will help consumers choose heart-healthy foods [press release]. July 9, 2003. Available at: http://www.hhs.gov/news/press/2003pres/20030709.html. Accessed November 16, 2004.

10. American Diabetes Association. Management of dyslipidemia in adults with diabetes. *Diabetes Care* 2003;26(suppl 1):S83-S86.

11. Balkau B, Shipley M, Jarrett RJ, et al: High blood glucose concentration is a risk factor for mortality in middle-aged nondiabetic men. *Diabetes Care* 1998;21:360-367.

12. Antonicelli R, Gesuita R, Boemi M, et al: Random fasting hyperglycemia as cardiovascular risk factor in the elderly: a 6-year longitudinal study. *Clin Cardiol* 2001;24:341-344.

13. Intensive blood-glucose control with sulphonylureas or insulin compared with conventional treatment and risk or complica-

tions in patients with type 2 diabetes (UKPDS 33). UK Prospective Diabetes Study (UKPDS) Group. *Lancet* 1998;352:837-853.

14. Reaven GM: Multiple CHD risk factors in type 2 diabetes: beyond hyperglycaemia. *Diabetes Obes Metab* 2002;4(suppl 1): S13-S18.

15. Action to Control Cardiovascular Risk in Diabetes (ACCORD): ClinicalTrials.gov. A service of the National Institutes of Health. Web site. Accessed October 7, 2004.

16. Kris-Etherton PM: Monounsaturated fatty acids and risk of cardiovascular disease. American Heart Association Nutrition Committee. *Circulation* 1999;100:1253-1258.

17. Kris-Etherton PM, Lichtenstein AH, Howard BV, et al, for the Nutrition Committee of the American Heart Association Council on Nutrition, Physical Activity, and Metabolism: Antioxidant vitamin supplements and cardiovascular disease. *Circulation* 2004;110:637-641.

18. IFST (Institute of Food Science and Technology [UK]) current hot topics: Phytosterol esters (plant sterol and stanol esters). Web site. Available at: http://www.ifst.org/hottop29.htm. Accessed June 15, 2001.

19. Nguyen TT: The cholesterol-lowering action of plant stanol esters. *J Nutr* 1999;129:2109-2112.

20. Miettinen TA, Puska P, Gylling H, et al: Reduction of serum cholesterol with sitostanol-ester margarine in a mildly hypercholesterolemic population. *N Engl J Med* 1995;333:1308-1312.

21. Gylling H, Radhakrishnan R, Miettinen TA: Reduction of serum cholesterol in postmenopausal women with previous myocardial infarction and cholesterol malabsorption induced by dietary sitostanol ester margarine. *Circulation* 1997;96:4226-4231.

22. FDA authorizes new coronary heart disease health claim for plant sterol and plant stanol esters. FDA Talk Paper. Food and Drug Administration. Office of Public Affairs. September 5, 2000. Available at: http://www.fda.gov/bbs/topics/ANSWERS/ANS01033.html. Accessed October 7, 2004.

23. Frequently Asked Questions. Does Benecol contain trans fatty acids? Benecol Professional Resource [Web site]. Available at: http://www.benecolphysicians.com. Accessed July 2, 2004.

24. Taylor CL: FDA letter regarding enforcement discretion with respect to expanded use of an interim health claim rule about plant

sterol/stanol esters and reduced risk of coronary heart disease [letter]. U.S. Food and Drug Administration/Center for Food Safety and Applied Nutrition/Office of Nutritional Products, Labeling, and Dietary Supplements. February 14, 2003. Available at: http://vm.cfsan.fda.gov/~dms/ds-ltr30.html. Accessed October 8, 2004.

25. Berger A, Jones PJ, Abumweis SS: Plant sterols: factors affecting their efficacy and safety as functional food ingredients. Lipids in Health and Disease 2004;3:5. pp 1-19. BioMed Central Open Access. Published 7 April 2004. Available at: http://www.pubmedcentral.nih.gov/articlerender.fcgi?artid=419367. Accessed October 8, 2004.

26. Lichtenstein AH, Deckelbaum RJ: AHA Science Advisory. Stanol/sterol ester-containing foods and blood cholesterol levels. A statement for healthcare professionals from the Nutrition Committee of the Council on Nutrition, Physical Activity, and Metabolism of the American Heart Association. *Circulation* 2001;103:1177-1179.

27. Harris WS: N-3 fatty acids and serum lipoproteins: human studies. *Am J Clin Nutr* 1997;65(5 suppl):1645S-1654S.

28. Agren JJ, Vaisanen S, Hannien O, et al: Hemostatic factors and platelet aggregation after a fish-enriched diet or fish oil or docosahexaenoic acid supplementation. *Prostagland Leukot Essent Fatty Acids* 1997;57:419-421.

29. Mori TA, Beilin LJ, Burke V, et al: Interactions between dietary fat, fish, and fish oils and their effects on platelet function in men at risk of cardiovascular disease. *Arterioscler Thromb Vasc Biol* 1997;17:279-286.

30. Hu FB, Stampfer MJ, Manson JE, et al: Dietary intake of α-linolenic acid and risk of fatal ischemic heart disease among women. *Am J Clin Nutr* 1999;69:890-897.

31. Albert CM, Hennekens CH, O'Donnell CJ, et al: Fish consumption and risk of sudden cardiac death. *JAMA* 1998;279:23-28.

32. de Lorgeril M, Salen P, Martin JL, et al: Mediterranean diet, traditional risk factors, and the rate of cardiovascular complications after myocardial infarction; final report of the Lyon Diet Heart Study. *Circulation* 1999;99:779-785.

33. Singh RB, Niaz MA, Sharma JP, et al: Randomized, double-blind, placebo-controlled trial of fish oil and mustard oil in patients with suspected acute myocardial infarction: the Indian experiment of infarct survival. *Cardiovasc Drugs Ther* 1997;11:485-491.

34. Von Schacky C, Angerer P, Kothny W, et al: The effect of dietary ω-3 fatty acids on coronary atherosclerosis: a randomized, double-blind, placebo-controlled trial. *Ann Intern Med* 1999; 130:554-562.

35. Dietary supplementation with n-3 polyunsaturated fatty acids and vitamin E after myocardial infarction: results of the GISSI-Prevenzione trial. GISSI-Prevenzione Investigators. *Lancet* 1999; 354:447-455.

36. Daviglus ML, Stamler J, Orencia AJ, et al: Fish consumption and the 30-year risk of fatal myocardial infarction. *N Engl J Med* 1997;336:1046-1058.

37. Dyerberg J: Coronary heart disease in Greenland Inuit: a paradox. Implications for western diet patterns. *Arctic Med Res* 1989; 48:47-54.

38. Brown AM, Castle J, Hebbachi AM, et al: Administration of n-3 fatty acids in the diets of rats or directly to hepatocyte cultures results in different effects on hepatocellular Apo B metabolism and secretion. *Arterioscler Thromb Vasc Biol* 1999;19: 106-114.

39. Wu X, Shang A, Jiang H, et al: Demonstration of biphasic effects of docosahexaenoic acid on apolipoprotein B secretion in Hep G2 cells. *Arterioscler Thromb Vasc Biol* 1997;17:3347-3355.

40. Byrne CD, Wang TW, Hales CN: Control of Hep G2-cell triacylglycerol and apolipoprotein B synthesis and secretion by polyunsaturated non-esterified fatty acids and insulin. *Biochem J* 1992;288(Pt 1):101-107.

41. Ranheim T, Gedde-Dahl A, Rustan AC, et al: Influence of eicosapentaenoic acid (20:5, n-3) on secretion of lipoproteins in CaCo-2 cells. *J Lipid Res* 1992;33:1281-1293.

42. Lang CA, Davis RA: Fish oil fatty acids impair VLDL assembly and/or secretion by cultured rat hepatocytes. *J Lipid Res* 1990;31:2079-2086.

43. Arrol S, Mackness MI, Durrington PN: The effects of fatty acids on apolipoprotein B secretion by human hepatoma cells (HEP G2). *Atherosclerosis* 2000;150:255-264.

44. Schoonjans K, Staels B, Auwerx J: Role of the peroxisome proliferator-activated receptor (PPAR) in mediating the effects of fibrates and fatty acids on gene expression. *J Lipid Res* 1996;37: 907-925.

45. Pakala R, Pakala R, Benedict CR: Thromboxane A_2 fails to induce proliferation of smooth muscle cells enriched with eicosapentaenoic acid and docosahexaenoic acid. *Prostaglandins Leukot Essent Fatty Acids* 1999;60:275-281.

46. Rao GH, Radha E, White JG: Effect of docosahexaenoic acid (DHA) on arachidonic acid metabolism and platelet function. *Biochem Biophys Res Commun* 1983;117:549-555.

47. Adan Y, Shibata K, Sato M, et al: Effects of docosahexaenoic and eicosapentaenoic acid on lipid metabolism, eicosanoid production, platelet aggregation and atherosclerosis in hypercholesterolemic rats. *Biosci Biotechnol Biochem* 1999;63:111-119.

48. Yamada N, Shimizu J, Wada M, et al: Changes in platelet aggregation and lipid metabolism in rats given dietary lipids containing different n-3 polyunsaturated fatty acids. *J Nutr Sci Vitaminol (Tokyo)* 1998;44:279-289.

49. Nieuwenhuys CM, Hornstra G: The effects of purified eicosapentaenoic and docosahexaenoic acids on arterial thrombosis tendency and platelet function in rats. *Biochim Biophys Acta* 1998;1390:313-322.

50. Stroh S, Elmadfa I: In vitro studies of the effect of different mixture proportions of ω-3 and ω-6 fatty acids on thrombocyte aggregation and thromboxane synthesis in human thrombocytes [in German]. *Z Ernahrungswiss* 1991;30:192-200.

51. Leaf A, Weber PC: Cardiovascular effects of n-3 fatty acids. *N Engl J Med* 1988;318:549-557.

52. Durrington PN, Bhatnagar D, Mackness MI, et al: An ω-3 polyunsaturated fatty acid concentrate administered for one year decreased triglycerides in simvastatin treated patients with coronary heart disease and persisting hypertriglyceridaemia. *Heart* 2001;85:544-548.

53. Sirtori CR, Crepaldi G, Manzato E, et al: One-year treatment with ethyl esters of n-3 fatty acids in patients with hypertriglyceridemia and glucose intolerance: reduced triglyceridemia, total cholesterol and increased HDL-C without glycemic alterations. *Atherosclerosis* 1998;137:419-427.

54. Nelson GJ, Schmidt PC, Bartolini GL, et al: The effect of dietary docosahexaenoic acid on plasma lipoproteins and tissue fatty acid composition in humans. *Lipids* 1997;32:1137-1146.

55. Minihand AM, Khan S, Leigh-Firbank EC, et al: Apo E polymorphism and fish oil supplementation in subjects with an atherogenic lipoprotein phenotype. *Arterioscler Thromb Vasc Biol* 2000;20:1990-1997.

56. Davidson MH, Maki KC, Kalcowski J, et al: Effects of docosahexaenoic acid on serum lipoproteins in patients with combined hyperlipidemia: a randomized, double-blind, placebo-controlled trial. *J Am Coll Nutr* 1997;16:236-243.

57. Hsu HC, Lee YT, Chen MF: Effect of n-3 fatty acids on the composition and binding properties of lipoproteins in hypertriglyceridemic patients. *Am J Clin Nutr* 2000;71:28-35.

58. Calabresi L, Donati D, Pazzucconi F, et al: Omacor in familial combined hyperlipidemia: effects on lipids and low density lipoprotein subclasses. *Atherosclerosis* 2000;148:387-396.

59. Stalenhoef AF, de Graaf J, Wittekoek ME, et al: The effect of concentrated n-3 fatty acids versus gemfibrozil on plasma lipoproteins, low density lipoprotein heterogeneity and oxidizability in patients with hypertriglyceridemia. *Atherosclerosis* 2000;153: 129-138.

60. Burr ML, Fehily Am, Gilbert JF, et al: Effects of changes in fat, fish, and fibre intakes on death and myocardial reinfarction: Diet and Reinfarction Trial (DART). *Lancet* 1989;2:757-761.

61. FDA announces qualified health claims for omega-3 fatty acids. FDA News. U.S. Food and Drug Administration. September 8, 2004. Available at: http://www.fda.gov/bbs/topics/news/2004/NEW01115.html. Accessed October 8, 2004.

62. Squires S: Omega-3 foods can put benefits on label, FDA says. September 9, 2004. *Washington Post*, page A04. Available at http://www.washingtonpost.com/wp-dyn/articles/A6548-2004Sep8.html. Accessed October 8, 2004.

63. Kris-Etherton PM, Harris WS, Appel LJ, for the Nutrition Committee: Fish consumption, fish oil, omega-3 fatty acids, and cardiovascular disease. *Circulation* 2002;106:2747-2757.

64. United States Department of Agriculture. Dietary Guidelines for Americans 2000. Web site. Available at: http://www.usda.gov/cnpp/Pubs/DG2000/?S=D. Accessed October 8, 2004.

65. Menotti A, Kromhout D, Blackburn H, et al: Food intake patterns and 25-year mortality from coronary heart disease: cross-cul-

tural correlations in the Seven Countries Study. The Seven Countries Study Research Group. *Eur J Epidemiol* 1999;15:507-515.

66. Keys A, Menotti A, Karvonen MJ, et al: The diet and 15-year death rate in the Seven Countries Study. *Am J Epidemiol* 1986;124: 903-915.

67. Joshipura KJ, Hu FB, Manson JE, et al: The effect of fruit and vegetable intake on risk for coronary heart disease. *Ann Intern Med* 2001;134:1106-1114.

68. Millen BE, Quatromoni PA: Nutritional research within the Framingham Heart Study. *J Nutr Health Aging* 2001;5:139-143.

69. Renaud S, deLorgeril M, Delaye J, et al: Cretan Mediterranean diet for prevention of coronary heart disease. *Am J Clin Nutr* 1995;61(6 suppl):1360S-1367S.

70. Hellenius ML, deFaire U, Berglund B, et al: Diet and exercise are equally effective in reducing risk for cardiovascular disease. Results of a randomized controlled study in men with slightly to moderately raised cardiovascular risk factors. *Atherosclerosis* 1993;103:81-91.

71. Yu-Poth S, Zhao G, Etherton T, et al: Effects of the National Cholesterol Education Program's Step I and Step II dietary intervention programs on cardiovascular disease risk factors: a meta-analysis. *Am J Clin Nutr* 1999;69:632-646.

72. Schaefer EJ, Lichtenstein AH, Lamon-Fava S, et al: Efficacy of a National Cholesterol Education Program Step 2 diet in normolipidemic and hypercholesterolemic middle-aged and elderly men and women. *Arterioscler Thromb Vasc Biol* 1995;15: 1079-1085.

73. Obarzanek E, Sacks FM, Vollmer WM, et al: Effects on blood lipids of a blood pressure-lowering diet: the Dietary Approaches to Stop Hypertension (DASH) trial. *Am J Clin Nutr* 2001;74:80-89.

74. McLaughlin T, Abbasi F, Kim HS, et al: Relationship between insulin resistance, weight loss, and coronary heart disease risk in healthy, obese women. *Metabolism* 2001;50:795-800.

75. Brown AA, Hu FB: Dietary modulation of endothelial function: implications for cardiovascular disease. *Am J Clin Nutr* 2001; 73:673-686.

76. Ridker PM, Hennekens CH, Buring JE, et al: C-reactive protein and other markers of inflammation in the prediction of cardiovascular disease in women. *N Engl J Med* 2000;342:836-843.

77. Heilbronn LK, Noakes M, Clifton PM: Energy restriction and weight loss on very-low-fat diets reduce C-reactive protein concentrations in obese, healthy women. *Arterioscler Thromb Vasc Biol* 2001;21:968-970.

78. Kromhout D: Serum cholesterol in cross-cultural perspective. The Seven Countries Study. *Acta Cardiol* 1999;54:155-158.

79. Renaud S, deLorgeril M: Dietary lipids and their relation to ischaemic heart disease: from epidemiology to prevention. *J Intern Med Suppl* 1989;225:39-46.

80. Steinberg D, Gotto AM: Preventing coronary artery disease by lowering cholesterol levels. Fifty years from bench to bedside. *JAMA* 1999;282:2043-2050.

81. Wallace AJ, Mann JI, Sutherland WH, et al: Variants in the cholesterol ester transfer protein and lipoprotein lipase genes are predictors of plasma cholesterol response to dietary change. *Atherosclerosis* 2000;152:327-336.

82. Ye SQ, Kwiterovich PO Jr: Influence of genetic polymorphisms on responsiveness to dietary fat and cholesterol. *Am J Clin Nutr* 2000;72(5 suppl):1275S-1284S.

83. Ordovas JM, Schaefer EJ: Treatment of dyslipidemia: genetic interactions with diet and drug therapy. *Curr Atheroscler Rep* 1999; 1:16-23.

84. de Lorgeril M, Salen P, Martin JL, et al: Mediterranean diet, traditional risk factors, and the rate of cardiovascular complications after myocardial infarction. Final report of the Lyon Diet Heart Study. *Circulation* 1999;99:779-785.

85. Kris-Etherton P, Eckel RH, Howard BV, et al: Lyon Diet Heart Study. Benefits of a Mediterranean-style, National Cholesterol Education Program/American Heart Association Step I dietary pattern on cardiovascular disease. *Circulation* 2001;103:1823-1825.

86. Caggiula AW, Watson JE, Kuller LH, et al: Effect of the Step I diet in community office practices. Cholesterol-Lowering Intervention Program. *Arch Intern Med* 1996;156:1205-1213.

87. Rhodes KS, Bookstein LC, Aaronson LS, et al: Intensive nutrition counseling enhances outcomes of National Cholesterol Education Program dietary therapy. *J Am Diet Assoc* 1996;96:1003-1010.

88. Millen BE, Quatromoni PA, Franz MM, et al: Population nutrient intake approaches dietary recommendations: 1991 to 1995 Framingham Nutrition Studies. *J Am Diet Assoc* 1997;97:742-749.

89. Yarzebski J, Spencer F, Goldberg RJ, et al: Temporal trends (1986-1997) in cholesterol level assessment and management practices in patients with acute myocardial infarction: a population-based perspective. *Arch Intern Med* 2001;161:1521-1528.

90. Boekeloo BO, Becker DM, LeBailly A, et al: Cholesterol management in patients hospitalized for coronary heart disease. *Am J Prev Med* 1988;4:128-132.

91. Hueston WJ, Spencer E, Kuehn R: Differences in the frequency of cholesterol screening in patients with Medicaid compared with private insurance. *Arch Fam Med* 1995;4:331-334.

92. Harnick DJ, Cohen JL, Schechter CB, et al: Effects of practice setting on quality of lipid-lowering management in patients with coronary artery disease. *Am J Cardiol* 1998;81:1416-1420.

93. Levin SJ, Ornstein SM: Management of hypercholesterolemia in a family practice setting. *J Fam Pract* 1990;31:613-617.

94. Smith SC: Clinical treatment of dyslipidemia: practice patterns and missed opportunities. *Am J Cardiol* 2000;86(12 suppl 1):62-65.

95. Meigs JB, Stafford RS: Cardiovascular disease prevention practices by U.S. physicians for patients with diabetes. *J Gen Intern Med* 2000;15:220-228.

96. Solberg LI, Kottke TE, Brekke ML: Variation in clinical preventive services. *Eff Clin Pract* 2001;4:121-126.

97. Glanz K, Gilboy MB: Physicians, preventive care, and applied nutrition: selected literature. *Acad Med* 1992;67:776-781.

98. Ammerman AS, DeVellis RF, Carey TS, et al: Physician-based diet counseling for cholesterol reduction: current practices, determinants, and strategies for improvement. *Prev Med* 1993;22:96-109.

99. Secker-Walker RH, Morrow AL, Kresnow M, et al: Family physicians' attitudes about dietary advice. *Fam Pract Res J* 1991; 11:161-170.

100. Evans AT, Rogers LQ, Peden JG Jr, et al: Teaching dietary counseling skills to residents: patient and physician outcomes. The CADRE Study Group. *Am J Prev Med* 1996;12:259-265.

101. Legato MJ, Padus E, Slaughter E: Women's perceptions of their general health, with special reference to their risk of coronary artery disease: results of a national telephone survey. *J Womens Health* 1997;6:189-198.

102. Gault R, Yeater RA, Ullrich IH: West Virginia physicians: cardiovascular risk factors, lifestyles and prescribing habits. *W V Med J* 1994;90:364-366.

103. Williams CL, Bollella J, Wynder E: Preventive cardiology in primary care. *Atherosclerosis* 1994;108:S117-S126.

104. Superko HR, Desmond DA, de Santos VV, et al: Blood cholesterol treatment attitudes of community physicians: a major problem. *Am Heart J* 1988;116:849-855.

105. Tunis SR, Hayward RS, Wilson MC: Internists' attitudes about clinical practice guidelines. *Ann Intern Med* 1994;120:956-963.

106. Greenland P: Closing the treatment gap: in the community and at hospital discharge. *Am J Med* 1996;101:4A76S-4A78S.

107. Cooper R, Cutler J, Desvigne-Nickens P: Trends and disparities in coronary heart disease, stroke, and other cardiovascular diseases in the United States: findings of the National Conference on Cardiovascular Disease Prevention. *Circulation* 2000;102:3137-3147.

108. Thompson PD, Buchner D, Pina IL, et al: Exercise and physical activity in the prevention and treatment of atherosclerotic cardiovascular disease. A statement from the Council on Clinical Cardiology (Subcommittee on Exercise, Rehabilitation and Prevention) and the Council on Nutrition, Physical Activity, and Metabolism (Subcommittee on Physical Activity). *Circulation* 2003;107:3109-3116.

109. O'Connor GT, Hennekens CH, Willett WC, et al: Physical exercise and reduced risk of nonfatal myocardial infarction. *Am J Epidemiol* 1995;142:1147-1156.

110. Maines TY, Lavie CJ, Milani RV, et al: Effects of cardiac rehabilitation and exercise programs on exercise capacity, coronary factors, behavior, and quality of life in patients with coronary artery disease. *South Med J* 1997;90:43-49.

111. Lavie CJ, Milani RV: Effects of cardiac rehabilitation and exercise training programs in patients ≥ 75 years of age. *Am J Cardiol* 1996;78:675-677.

112. Dengel JL, Katzel LI, Goldberg AP: Effect of an American Heart Association diet, with or without weight loss, on lipids in obese middle-aged and older men. *Am J Clin Nutr* 1995;62:715-721.

113. Crouse SF, O'Brien BC, Grandjean PW, et al: Training intensity, blood lipids, and apolipoproteins in men with high cholesterol. *J Appl Physiol* 1997;82:270-277.

114. Pollock ML, Franklin BA, Balady GJ, et al: Resistance exercise in individuals with and without cardiovascular disease. Benefits, rationale, safety, and prescription. An advisory from the Committee on Exercise, Rehabilitation, and Prevention, Council on Clinical Cardiology, American Heart Association. *Circulation* 2000;101:828-833.

115. Ockene IS, Miller NH: Cigarette smoking, cardiovascular disease, and stroke. A statement for healthcare professionals from the American Heart Association Task Force on Risk Reduction. *Circulation* 1997;96:3243-3247.

116. Froom P, Melamed S, Benbasset J: Smoking cessation and weight gain. *J Fam Pract* 1998;46:460-464.

117. Goldberg IJ, Mosca L, Piano MR, et al: AHA Science Advisory: Wine and your heart: a science advisory for healthcare professionals from the Nutrition Committee, Council on Epidemiology and Prevention, and Council on Cardiovascular Nursing of the American Heart Association. *Circulation* 2001;103:472-475.

Therapeutic Options: Pharmacologic Interventions

Based primarily on evidence from a series of large, randomized, controlled trials, the 2001 guidelines of the National Cholesterol Education Program (NCEP), Adult Treatment Panel III (ATP III), state that the primary goal of cardiovascular risk-reduction therapy is to reduce the level of low-density lipoprotein cholesterol (LDL-C).[1] In 2004, an update to the guidelines was issued that examines the implications of five major clinical trials completed since the ATP III recommendations were published.[2] These trials of statin therapy and clinical end points examined issues that were not adequately addressed in earlier major investigations, including the treatment of underrepresented patient populations (eg, women, the elderly) and of patients with established coronary heart disease (CHD) and/or multiple risk factors who were not definite candidates for LDL-lowering therapy based on accepted guidelines. The update, while reinforcing LDL lowering as the primary target of therapy, also focuses on the concept of overall risk and endorses the use of a variety of risk-reduction strategies. Table 5-1 lists the principal features of the revised ATP III guidelines.

Although therapeutic lifestyle changes (TLC) are an essential part of any cardiovascular risk-reduction program, they are not always sufficient. For many patients, particularly those at high or moderately high risk, drug therapy is essential. When compared with the costs of cardiovascular disease (CVD), drug therapy can also be the most cost-effective strategy. Because poor dietary habits can offset the benefits of lipid-lowering therapy, all patients must continue TLC after drug treatment is initiated.

For patients who begin with TLC, ATP III recommends two follow-up visits at 6-week intervals for the physician to determine the LDL-C response (Chapter 4, Figure 4-1). If after 12 weeks of TLC, the patient's LDL-C level is still unacceptably high, drug therapy should be considered. The steps for initiating and monitoring drug therapy are shown in Figure 5-1.

Before drug therapy is initiated, any causes of secondary dyslipidemia should be identified. Secondary dyslipidemia may have several causes, including diabetes, hypothyroidism, obstructive liver disease, chronic renal failure, or drugs such as progestins, anabolic steroids, and corticosteroids (Chapter 3, Table 3-2). Once underlying causes of secondary dyslipidemia have been addressed, the goals for lipid-lowering therapy can be established according to the patient's risk category. Underlying conditions may also affect decisions regarding drug or dietary therapy. For example, patients with renal dysfunction require particular attention to protein and salt intake,[3] and are also particularly at risk for diabetes, a CHD risk equivalent.

Percentage Reduction in LDL: Treatment Implications

Recent clinical trial evidence indicates that each 1% decrease in LDL-C produces a corresponding reduction of approximately 1% in the relative risk for CHD. In the Heart Protection Study (HPS), this correlation was observed even in subjects with baseline LDL-C levels <100 mg/dL.[4] At doses used in these trials (termed 'standard doses'), statin therapy can be expected to decrease LDL-C levels by 30% to 40%. Over a 5-year period, a similar percentage reduction in CHD risk is expected.

Table 5-1: Principal Features of the Updated ATP III Guidelines

Primary goal of therapy	• Lower LDL-C levels
Secondary goal of therapy	• Reduce non-HDL-C (when TG level is 200-499 mg/dL). Goal is 30 mg/dL higher than LDL-C goal.
Therapeutic lifestyle changes	• Modify lifestyle-related risk factors (all patients, particularly those at high or moderately high risk)
Patient classification	• High risk: CHD/CHD risk equivalents (including ≥2 risk factors with 10-yr CHD risk >20%*) • Moderately high risk: ≥2 risk factors (10-yr CHD risk 10% to 20%*) • Moderate risk: ≥2 risk factors (10-yr CHD risk <10%) • Lower risk: 0-1 risk factor**

ATP = Adult Treatment Panel; CHD = coronary heart disease; HDL-C = high-density lipoprotein cholesterol; LDL-C = low-density lipoprotein cholesterol; TG = triglycerides; TLC = therapeutic lifestyle changes

The ATP III guidelines emphasize therapy to achieve specific LDL-C goals, rather than the reduction of LDL-C by a given percentage. However, the 2004 update, while not recommending a definite proportional reduction in LDL-

Guidelines for initiating drug therapy (mg/dL)	• CHD/CHD risk equivalents – if baseline LDL-C is ≥100 mg/dL, simultaneously initiate TLC and drug therapy to reach lipid goal – if patient has high TG or low HDL-C, consider combining a fibrate or nicotinic acid with LDL-lowering therapy • ≥2 risk factors (moderately high risk, moderate risk) – consider drug options if TLC does not achieve goal within 3 mo • 0-1 risk factor – consider drug options if an adequate trial of TLC does not reduce LDL-C to <160 mg/dL and there is a single, severe risk factor or other factor favoring drug use

* See Chapter 3 for an explanation of risk equivalents and risk factors.

** 10-yr risk assessment not necessary (almost all people with 0-1 risk factor have a 10-yr risk <10%).

(continued on next page)

C, considers it reasonable to use drug doses that will achieve at least a 30% to 40% LDL decrease in patients at high or moderately high risk. Patients with relatively low baseline LDL-C levels may be able to reach the minimum goal (<100

Table 5-1: Principal Features of the Updated ATP III Guidelines *(continued)*

LDL-C cut points for initiating drug therapy (mg/dL)	• CHD/CHD risk equivalents*: ≥100 mg/dL (<100 mg/dL: consider drug options) • ≥2 risk factors: – 10-yr risk 10% to 20%: ≥130 mg/dL (100-129 mg/dL: consider drug options) – 10-yr risk <10%: ≥160 mg/dL • 0-1 risk factor: ≥190 mg/dL (160-189 mg/dL: consider drug options)

ATP = Adult Treatment Panel
CHD = coronary heart disease
LDL-C = low-density lipoprotein cholesterol

mg/dL) or the more rigorous optional goal (<70 mg/dL) either with standard-dose statin therapy or with a lower-dose statin combined with another product (eg, ezetimibe [Zetia®], a bile-acid sequestrant, plant stanols/sterols). However, those with higher LDL-C levels may require more intensive therapy (within the bounds of safety and tolerability) to achieve the minimum goal.

Based on the HPS data, some experts support the use of statin therapy in all high-risk patients with a baseline LDL-C <100 in order to reach a level well below the current minimum goal. However, ATP III refrains from making such a recommendation because of limitations in the HPS analysis. The results of ongoing clinical trials (eg, Treating to New Targets [TNT], Study of the Effectiveness of Additional Reductions in Cholesterol and Homocysteine [SEARCH], Incremen-

LDL-C goals (mg/dL)	• CHD/CHD risk equivalent: <100 mg/dL (<70 mg/dL: optional goal) • ≥2 risk factors: – 10-yr risk 10% to 20%: <130 mg/dL (<100 mg/dL: optional goal) – 10-yr risk <10%: <130 mg/dL • 0-1 risk factor: <160 mg/dL

* See Chapter 3 for an explanation of risk equivalents and risk factors.

tal Decrease in Endpoints through Aggressive Lipid Lowering [IDEAL]) may provide additional support for an LDL-C goal well below 100 mg/dL in all high-risk patients. In some patients with very high baseline levels, an LDL goal of <100 mg/dL may be unachievable.

Risk Categories

CHD and CHD Risk Equivalents (High Risk)

The minimum LDL-C goal for all patients with CHD or CHD risk equivalents (see Chapter 3) is <100 mg/dL. For patients at very high risk, an LDL-C goal of <70 mg/dL is a therapeutic option. Very high risk is defined as the presence of established CVD plus (1) multiple major risk factors (particularly diabetes), (2) severe or poorly controlled risk factors (particularly cigarette smoking), (3) multiple risk fac-

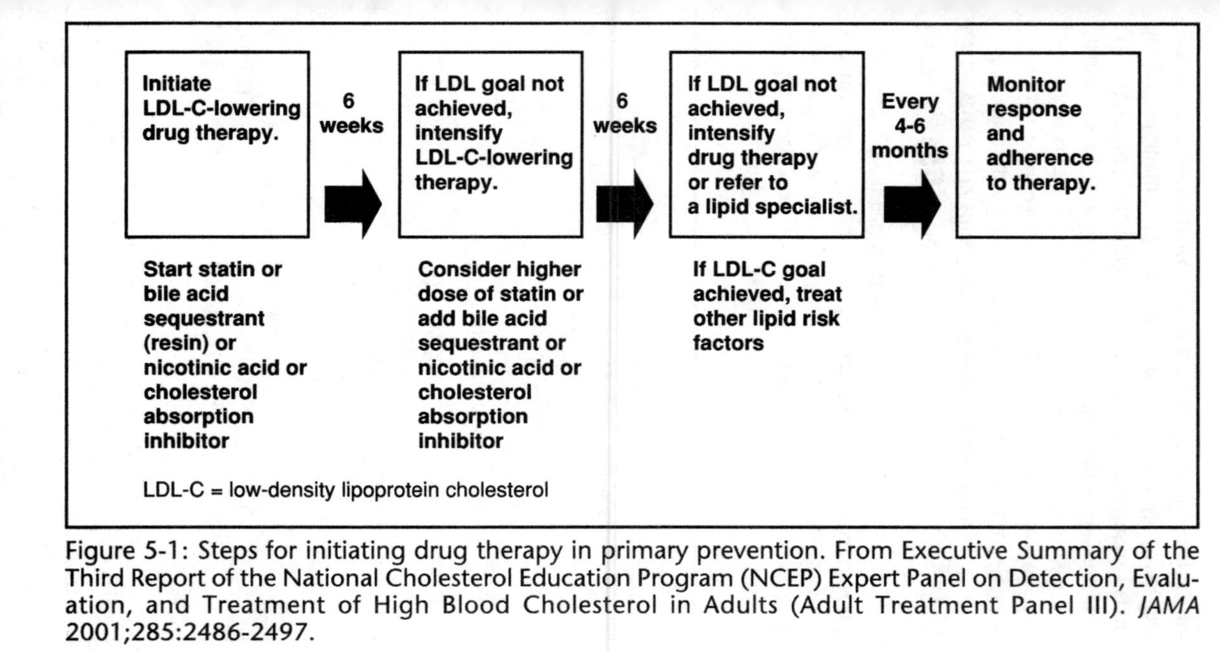

Figure 5-1: Steps for initiating drug therapy in primary prevention. From Executive Summary of the Third Report of the National Cholesterol Education Program (NCEP) Expert Panel on Detection, Evaluation, and Treatment of High Blood Cholesterol in Adults (Adult Treatment Panel III). *JAMA* 2001;285:2486-2497.

tors characteristic of the metabolic syndrome (particularly high triglycerides [TG] plus non-high-density lipoprotein cholesterol [non-HDL-C] ≥130 mg/dL with low HDL-C [<40 mg/dL]), or (4) an acute coronary syndrome (ACS).

If the baseline LDL-C level is ≥100 mg/dL, simultaneous initiation of lipid-lowering drug therapy and TLC is recommended. Weight reduction and increased physical activity are also the first-line option for treating TG elevations (≥150 mg/dL). In addition, the physician can consider combining a fibrate or nicotinic acid (Niacor®, Niaspan® Extended Release) with LDL-lowering therapy in patients with TG levels ≥200 mg/dL or low HDL-C. However, coadministration of a statin and a fibrate can increase the risk for certain adverse events, including myopathy. See Drug Therapy/Fibric Acid Derivatives (this chapter) for a summary of precautions.

Regardless of drug regimen, compliance with TLC must be monitored. If adherence is a problem, referral to a dietitian or nutritionist for medical nutrition therapy is recommended.

Hospitalized patients. When a patient is hospitalized for an ACS or a coronary procedure, lipids should be measured at the time of admission or within 24 hours. The results can be used to guide drug treatment decisions either during hospitalization or at discharge. Clinicians must remember, however, that LDL-C levels decline in the first few hours after a coronary event. This decline continues during the first 24 to 48 hours after the event, and levels may remain low for several weeks.

Chapter 1 examines clinical trials of lipid-lowering therapy in ACS patients, and Chapter 6 reviews treatment guidelines.

Multiple Risk Factors

Moderately high risk. For patients with ≥2 risk factors and an estimated 10-year CHD risk of 10% to 20%, the LDL-C cut point for considering lipid-lowering drug therapy is ≥130 mg/dL after a 3-month trial of lifestyle modification. In selected patients, the physician may choose to consider drug therapy at an LDL-C level of 100 to 129 mg/dL, with the option of reducing the LDL-C level to <100 mg/dL (based on results of the Anglo-Scandinavian Cardiac Outcomes Trial—Lipid Lowering Arm [ASCOT-LLA]).

Deciding when to initiate lipid-lowering therapy in moderately high-risk patients requires physicians to consider the limitations of the Framingham risk estimate. In some patients, the Framingham formula may underestimate CHD risk because it does not include certain major, metabolic, or emerging risk factors (eg, premature CHD in a first-degree relative; abdominal obesity, elevated TG levels, impaired fasting glucose; elevated levels of C-reactive protein or lipoprotein (a) [Lp(a)]) that can increase actual risk, possibly to a level >20%. Consequently, if overall risk is deemed higher than the Framingham estimate, the physician may decide to initiate lipid-lowering therapy at an LDL-C level of 100 to 129 mg/dL (at baseline or on lifestyle therapy), with a goal of <100 mg/dL, and may also choose to begin drug therapy simultaneously with TLC. Moderately high-risk patients with the metabolic syndrome or individual lifestyle-related risk factors are also candidates for TLC regardless of their LDL-C level. As with high-risk patients, the intensity of therapy in moderately high-risk patients should be sufficient (within the bounds of safety and tolerability) to achieve at least a 30% to 40% reduction in LDL-C level.

Moderate risk. For patients with ≥2 risk factors and a Framingham risk estimate of <10%, the LDL threshold for initiating lipid-lowering therapy is ≥160 mg/dL after a 3-month trial of TLC, and the LDL-C goal is <130 mg/dL. Once again, premature CHD in a first-degree relative, as well as metabolic and emerging risk factors, may mean that some patients have a CHD risk higher than that predicted by the Framingham score. Thus, global risk assessment and the use of clinical judgment are crucial in determining both the appropriate treatment regimen and when to initiate drug therapy.

Less Than Two Risk Factors

Generally, persons with 0 to 1 risk factor have a <10% risk of developing CHD within 10 years. Therefore, it is not necessary to calculate individual risk in these patients. Because they are considered to have long-term, but not short-term, risk, their LDL-C goal is <160 mg/dL. Therapeutic lifestyle changes are first-line therapy, and use of medication is generally not considered cost-effective. However, if the LDL-C level is 160 to 189 mg/dL after an adequate trial of TLC, the physician should use clinical judgment in determining whether drug therapy is appropriate. The presence of any of the following strengthens the case for drug therapy: a single severe risk factor (eg, cigarette smoking, poorly controlled hypertension, presence of familial hypercholesterolemia, strong family history of premature CHD, low HDL-C level); multiple life-habit risk factors (see Chapter 3); 10-year risk close to 10% (if measured); or LDL-C level ≥190 mg/dL. An LDL-C level ≥190 mg/dL after TLC provides particularly strong support for the cost-effectiveness of drug therapy in patients without clinical evidence of CHD. Once again, because some risk factors are not included in the Framingham scoring system (eg, parental history of premature CVD),[5] clinical judgment must determine when actual risk exceeds the Framingham estimate (see Appendix A, Case 6).

The Metabolic Syndrome

The metabolic syndrome, which increases the risk for CHD regardless of LDL-C level or risk category, must be addressed with lifestyle and, when appropriate, pharmacologic interventions as part of a comprehensive risk-reduction plan.

Defined as "a constellation of lipid and nonlipid risk factors of metabolic origin," the metabolic syndrome is diagnosed when three or more of the following are present: abdominal obesity (waist circumference: men >40 in, women >35 in); elevated TG (≥150 mg/dL); low HDL-C (men <40 mg/dL; women <50 mg/dL); high blood pressure (≥130/≥85 mm Hg); and elevated fasting plasma glucose (≥100 mg/dL).[6,7] The metabolic syndrome is closely associated with insulin resistance, which is promoted by excess fat, particularly abdominal obesity. Some individuals may also have a genetic predisposition to insulin resistance.

Weight reduction and increased physical activity are first-line therapies for the metabolic syndrome. In addition, the elevated TG and decreased HDL-C levels may require pharmacologic therapy with a fibrate or nicotinic acid.

Elevated Serum Triglycerides

Elevated TG levels (≥150 mg/dL) are an independent risk factor for CHD. While weight reduction and physical activity are first-line therapies for borderline high TG (150 to 199 mg/dL), drug therapy (together with TLC) is the preferred approach for TG levels ≥200 mg/dL (Table 5-2). The treatment strategy for elevated TG is determined by the cause of the elevation and its severity. The goal of all therapy is to achieve a TG level <150 mg/dL.

Because high TG levels are an independent CHD risk factor, it is thought that certain TG-rich lipoproteins may be atherogenic. These include partially degraded very-low-density lipoprotein (VLDL), known as *remnant lipoproteins*. VLDL-C is the most readily available measure of remnant lipoproteins in clinical practice, thus constituting a target of therapy. In persons with high TG levels (200 to 499 mg/dL), reducing the concentration of LDL-C + VLDL-C (ie, non-HDL-C) is a secondary therapeutic objective after achieving the LDL-C goal.

Non-HDL-C may be reduced by (1) intensifying LDL-C-lowering therapy, and (2) further decreasing VLDL-C levels by cautiously adding a fibrate or nicotinic acid to the LDL-lowering regimen. Because a VLDL-C of ≤30 mg/dL is considered normal, the non-HDL-C goal for each risk category is set at 30 mg/dL higher than the LDL-C goal (Table 5-3).[1] In high-risk patients with elevated TG levels, consideration can be given to combining a fibrate or nicotinic acid with an LDL-lowering drug at the start of therapy.[2]

For patients with TG levels ≥500 mg/dL, TG lowering takes precedence over LDL-C lowering because of the risk for acute pancreatitis. In addition to a fibrate or nicotinic acid in most cases, therapy includes a low fat intake (≤15%

Table 5-2: Triglyceride Levels and Drug Interventions

Classification (mg/dL)	Drug Therapy	TG Goal (mg/dL)
Normal	None	<150
Borderline-high (150-199)	If possible, modify dosage of drugs that cause secondary TG elevations: • corticosteroids • estrogens • retinoids • β-adrenergic blockers • other	<150
High (200-499)	All of the above *plus* • intensification of LDL-C-lowering therapy to reduce non-HDL-C (ie, LDL-C + VLDL-C, or TC-HDL-C) *or* • cautious addition of nicotinic acid or fibrate to reduce VLDL-C level	<150
Very high (≥500)	All of the above	<150

HDL-C = high-density lipoprotein cholesterol; LDL-C = low-density lipoprotein cholesterol; TG = triglycerides; VLDL-C = very-low-density lipoprotein cholesterol

Table 5-3: Non-HDL-C and LDL-C Goals for Patients With High TG Levels

Risk Category	LDL-C Goal (mg/dL)	Non-HDL-C Goal (mg/dL)
CHD and CHD risk equivalent	<100 (<70: optional goal)	<130 (<100: optional goal)
Moderately high risk	<130 (<100: optional goal)	<160 (<130: optional goal)
Moderate risk	<130	<160
0-1 risk factor	<160	<190

of daily calories), weight reduction, and increased physical activity. Once TG levels are <500 mg/dL, LDL-C becomes the focus of attention.

Low HDL-C

In the ATP III guidelines, an HDL-C level <40 mg/dL is a major risk factor, and an HDL-C level ≥60 mg/dL is a 'negative risk factor' (ie, its presence removes one risk factor from the total number). Nevertheless, the panel does not recommend a numeric goal for increasing HDL-C levels because the clinical trial data are insufficient to establish a goal and currently available drugs lack a robust ability to raise HDL-C levels.

Low HDL-C is caused by many factors associated with insulin resistance (elevated TG, excess weight, physical inactivity, diabetes mellitus), as well as by cigarette smoking and high carbohydrate consumption (>60% of caloric intake).

Low HDL-C may also be secondary to specific drugs (eg, β-blockers, anabolic steroids, progestational agents, retinoids) (see Chapter 3, Table 3-2). Therefore, a complete medical history is essential. Weight reduction and increased physical activity can help raise HDL-C in all patients. If low HDL-C is accompanied by a high TG level, then non-HDL-C becomes a secondary target of therapy once the LDL-C goal has been achieved.[1] However, physicians can consider combining an HDL-raising drug (ie, a fibrate or nicotinic acid) with an LDL-lowering drug at the start of therapy in high-risk patients with low HDL-C levels.[2] See Chapter 3 (Other Issues in Risk Assessment) for a further discussion of HDL-C.

Drug Therapy

The decision to initiate drug therapy is based on several factors, including assessment of the risk-benefit ratio, cost, and compliance. Because drug therapy may represent a lengthy or lifelong commitment for the patient, clinical judgment is essential to ensure that the patient's best interests are represented. A variety of lipid-modifying agents is available, with varying costs and differing side effect profiles.[8] These agents make it possible to tailor drug treatment to specific dyslipidemias (Tables 5-4 and 5-5). A profile of these drugs is provided in Table 5-6.

Treating Isolated Elevations of LDL-C

Bile Acid Sequestrants (Resins)

The bile acid sequestrants (resins) have been in use as lipid-regulating agents for more than 3 decades, and a large body of clinical experience has accumulated regarding the lipid-lowering efficacy, tolerability, and impact on clinical end points of these drugs.[9] Cholestyramine (LoCholest®, Questran®, Prevalite®), colestipol (Colestid®), and colesevelam (WelChol®) are the three available resins. Colesevelam, which was approved by the US Food and Drug Administration in 2000, is the newest of these agents. They differ in chemical structure (Figure 5-2) but share a common mechanism of ac-

Table 5-4: Lipid-Modifying Effects of Major Available Lipid-Regulating Drug Classes

	LDL-C	HDL-C	TG
Bile acid sequestrants	↓15%-30%	↑5%	←→
HMG-CoA reductase inhibitors (statins)*	↓20%-63%	↑5%-15%	↓10%-37%
Fibric acid derivatives	←→, ↓10%-15%	↑5%-20%	↓20%-50%
Cholesterol absorption inhibitor	↓**	↑***	↓***
Nicotinic acid	↓5%-25%	↑15%-35%	↓20%-50%

* An agent combining a statin and a cholesterol absorption inhibitor is now available.
** 18% as monotherapy, 39% to 56% with statin
*** Limited data available
↓, decrease; ↑, increase; ←→, variable

tion. The introduction of agents (eg, statins) that achieve greater cholesterol reductions and are generally well tolerated has led to the decline in the use of resins as monotherapy.

Mechanism of action. The chief mechanism of action of the bile acid sequestrants is to interrupt the normal recirculation of the cholesterol-rich bile acid pool.[8] Bile acids are synthesized from cholesterol in the liver. Under normal physiologic conditions, approximately 3% of the bile acid pool is excreted; the remainder undergoes enterohepatic recircula-

Table 5-5: Drug Selection Based on Lipid Fractions

	Single Drug	Combination Drug
Elevated LDL-C and TG <200 mg/dL	Bile-acid sequestrant (resin) HMG-CoA reductase inhibitor (statin) Ezetimibe Nicotinic acid	Resin + statin Statin + nicotinic acid* Statin + ezetimibe** Resin + nicotinic acid
Elevated LDL-C and TG 200-499 mg/dL	Nicotinic acid Statin Fibric acid derivative (fibrate)	Nicotinic acid + statin* Statin + fibrate† Nicotinic acid + resin Nicotinic acid + fibrate

* Possible increased risk for myopathy and hepatitis.
** Possible increased risk for transaminase elevations; contraindicated in patients with liver disease.
† Increased risk for myopathy. Use with caution.
LDL-C = low-density lipoprotein cholesterol; TG = triglycerides

tion after absorption in the distal ileum. Resins bind to bile acids in the intestine, preventing their recirculation back to the liver and increasing their excretion twofold to tenfold. Depletion of the bile acid pool via excretion leads to an increase in hepatic synthesis of bile acids, which reduces intrahepatic cholesterol. In addition to removing bile acids, resin therapy may also reduce the absorption of dietary cholesterol, but this effect is not considered quantitatively significant in long-term therapy.

Table 5-6: Available Lipid-Modifying Agents*	
	Bile Acid Sequestrants
Available drugs	cholestyramine, colestipol, colesevelam
Chief use	Lower LDL-C
Selected contraindications	Biliary obstruction Familial dysbetalipoproteinemia; TG >500 mg/dL

HDL-C = high-density lipoprotein cholesterol;
LDL-C = low-density lipoprotein cholesterol;
TG = triglycerides
*Consult complete prescribing information

By interrupting the enterohepatic circulation of bile acids, increasing their fecal excretion, and decreasing intrahepatic cholesterol, resins cause an upregulation of the LDL receptor, which allows for recognition, binding, and removal

Nicotinic Acid	HMG-CoA Reductase Inhibitors
Crystalline nicotinic acid, sustained-release (or slow release) nicotinic acid**	atorvastatin, fluvastatin, lovastatin,** pravastatin, rosuvastatin, simvastatin***
Useful in most lipid and lipoprotein abnormalities; particularly effective in raising HDL-C; also reduces Lp(a)	Lower LDL-C
Chronic liver disease Arterial bleeding Peptic ulcer disease	Active or chronic liver disease; unexplained persistent elevations in serum transaminase levels; pregnancy; renal dysfunction

**A combination agent (extended-release nicotinic acid/lovastatin) is also available (see discussion under HMG-CoA Reductase Inhibitors).
***A combination agent (ezetimibe/simvastatin) is also available (see discussion under Cholesterol Absorption Inhibitors).

(continued on next page)

from the plasma compartment of lipoproteins that contain surface apo B and apo E. Therefore, resins reduce cholesterol via a dual mechanism of action that involves increased plasma clearance of cholesterol-rich lipoproteins coupled

Table 5-6: Available Lipid-Modifying Agents* *(continued)*

	Bile Acid Sequestrants
Precautionary information	Can worsen hypertriglyceridemia (TG >200 mg/dL)
Major side/adverse effects	Gastric distress (eg, constipation, bloating, flatulence, nausea); decreased absorption of other drugs
Usual daily dose	cholestyramine: 4-16 g colestipol: 5-30 g (granules); 2-16 g (tablets) colesevelam: 2,500-3,750 mg (depending on monotherapy or combination therapy)

*Consult complete prescribing information

Nicotinic Acid	HMG-CoA Reductase Inhibitors
Can worsen glucose control; use cautiously in type 2 diabetes, severe gout, hyperuricemia	Concomitant use of certain drugs, including cyclosporine, gemfibrozil, nicotinic acid, warfarin; concomitant use of azole antifungals or macrolide antibiotics (atorvastatin, lovastatin, simvastatin); hematuria/proteinuria (tubular) with rosuvastatin 40 mg (generally transient, not associated with worsening renal function)
Flushing; hyperglycemia; hyperuricemia or gout; hepatotoxicity, especially for sustained-release form; gastrointestinal upset; headache; itching	Myopathy, transaminase elevations, headache, gastrointestinal upset, fatigue, flu-like symptoms
Crystalline nicotinic acid: 2-4 g Sustained-release nicotinic acid: 1-2 g	atorvastatin: 10-80 mg fluvastatin: 20-80 mg (80 mg available as XL only) lovastatin: 10-80 mg pravastatin: 10-80 mg rosuvastatin: 5-40 mg simvastatin: 5-80 mg

(continued on next page)

Table 5-6: Available Lipid-Modifying Agents* *(continued)*

	Bile Acid Sequestrants
Maximum daily dose	cholestyramine: 24 g colestipol: 30 g (granules); 16 g (tablets) colesevelam: 4,375 mg
Available preparations	cholestyramine • 9-g packets (4 g drug) • 378 g bulk cholestyramine 'light' • 5-g packets or 5.5-g packets (4 g drug) • 210 g bulk or 231 g bulk colestipol • 5-g packets or 7.5-g packets (flavored) (5 g drug) • 300 g bulk or 500 g bulk • 1-g tablets colesevelam • 625-mg tablets

*Consult complete prescribing information

with an interruption in the enterohepatic circulation of bile acids. The long-term efficacy of resin therapy is blunted by a secondary stimulation of 3-hydroxy-3-methylglutaryl coenzyme A (HMG-CoA) reductase activity in the liver, which increases cholesterol synthesis and returns cholesterol levels toward baseline. For this reason, resins are often coad-

Nicotinic Acid	HMG-CoA Reductase Inhibitors
Crystalline nicotinic acid: 4 g Sustained-release nicotinic acid: 2 g	atorvastatin: 80 mg fluvastatin: 40 mg; 80 mg XL lovastatin: 80 mg pravastatin: 80 mg rosuvastatin: 40 mg simvastatin: 80 mg
Crystalline nicotinic acid: 500-mg tablet Sustained-release nicotinic acid: 500, 750, 1,000 mg tablets	atorvastatin: 10, 20, 40, 80 mg tablets fluvastatin: 20, 40 mg tablets; 80 mg XL lovastatin: 10, 20, 40 mg tablets pravastatin: 10, 20, 40, 80 mg tablets rosuvastatin: 5, 10, 20, 40 mg tablets simvastatin: 5, 10, 20, 40, 80 mg tablets

(continued on next page)

ministered with statins, whose principal mechanism of action is to inhibit HMG-CoA reductase.

Dosage and efficacy. Cholestyramine and colestipol are administered as a powdered resin or as granules that may be mixed with liquids or combined with food to increase palatability. Colestipol is also available in 1-gram tablets, and

Table 5-6: Available Lipid-Modifying Agents* *(continued)*

	Fibric Acid Derivatives
Available drugs	Gemfibrozil Fenofibrate
Chief use	Lower TG and raise HDL-C Variable effect on LDL-C
Selected contra-indications	Hepatic or renal dysfunction Primary biliary cirrhosis Preexisting gallbladder disease
Precautionary information	Hypersensitivity to these drugs
Selected major side/adverse effects	Gastrointestinal complaints may increase lithogenicity of bile and risk for developing cholesterol gallstones may potentiate effect of oral anticoagulants; in combination with an HMG-CoA reductase inhibitor, can increase the risk for muscle toxicity (including rhabdomyolysis)
Usual daily dose	Gemfibrozil: 1,200 mg (600 mg b.i.d. before breakfast and dinner) Fenofibrate: 54 mg-160 mg with food
Maximum daily dose	Gemfibrozil: 1,200 mg (600 mg b.i.d. before meals) Fenofibrate: 160 mg (with food)
Available preparations	Gemfibrozil: 600 mg tablets Fenofibrate: 54, 160 mg tablets

	Cholesterol Absorption Inhibitors
Available drugs	Ezetimibe***
Chief use	Lower TC, LDL-C, apo B (monotherapy or in combination with a statin)
Selected contra-indications	Active liver disease; unexplained, persistent elevations in serum transaminases
Precautionary information	Not recommended in patients with hepatic insufficiency; efficacy may be reduced in combination with a resin; coadministration with a fibrate not recommended until use in patients is studied; careful monitoring needed in patients taking cyclosporine
Selected major side/adverse effects	Slightly increased frequency of transaminase elevations in combination with a statin (vs statin alone)
Usual daily dose	10 mg/d
Maximum daily dose	10 mg/d
Available preparations	10 mg tablets

*Consult complete prescribing information
***A combination agent (ezetimibe/simvastatin) is also available (see discussion under Cholesterol Absorption Inhibitors).

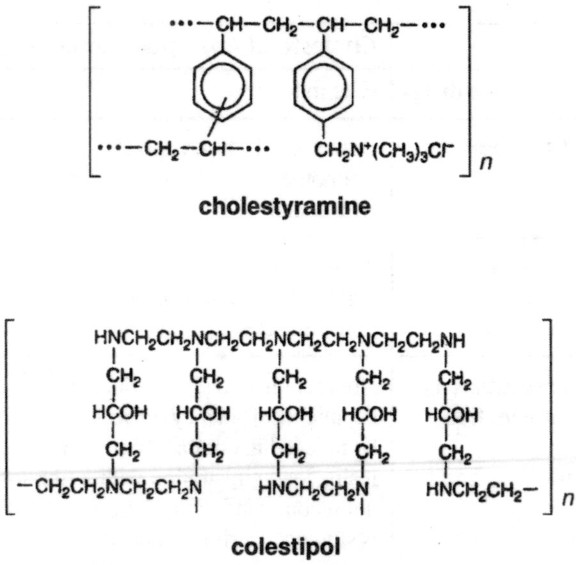

cholestyramine

colestipol

Figure 5-2: Structures of the bile acid sequestrants.

colesevelam is available in 625-mg tablets. Tablet dosing may contribute to ease of administration. Compliance with resin therapy may be improved if the initial dose is low and then slowly increased to the maximum level. The predominant therapeutic effect of resin therapy is the lowering of total cholesterol (TC) and LDL-C, and the expected reductions in LDL-C with monotherapy range from 15% to 30%. High-density lipoprotein cholesterol has been reported to increase by up to 5%, although the mechanism underlying this increase is not clear. The effects on TG are variable, and an increase in VLDL-C may occur after the administration of

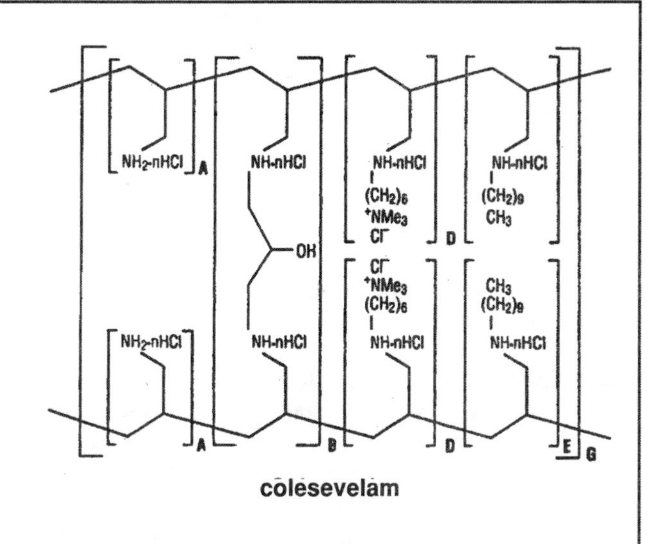

From Davidson et al, *Arch Intern Med* 1999;159:1893-1900.

resins. The risk for elevations in TG concentrations is increased when the pretreatment TG level is already elevated. The precise mechanism for the resin-induced increase in TG levels has not been established. The resins have no demonstrable effect on chylomicrons.

Side effects and drug interactions. Resin therapy is attractive because of its nonsystemic mechanism of action. However, the resins are often poorly tolerated, and achieving maximum dosage levels is difficult because of adverse gastrointestinal effects. These include bloating, nonspecific abdominal pain, and gastroesophageal reflux. Constipation is common and may be disabling in elderly patients, although the impact of resin therapy on bowel function may be ameliorated by increasing fluid and fiber intake or by using stool softeners.

Because they are anion exchange resins, the bile acid sequestrants may result in nonspecific binding of a large number of coadministered drugs, such as thyroxin, digoxin (Lanoxin®), warfarin (Coumadin®), diuretics, β-blockers, and a variety of antibiotics. The dosing of these drugs 1 hour before or 4 to 6 hours after ingestion of bile acid resins may minimize nonspecific binding and reduced absorption of concomitant medications.

HMG-CoA Reductase Inhibitors

The HMG-CoA reductase inhibitors (statins) were introduced into clinical practice more than 15 years ago and marked a significant advance in the pharmacologic management of dyslipidemias.[10] The six available statins are atorvastatin (Lipitor®), fluvastatin (Lescol®), lovastatin (Mevacor®), pravastatin (Pravachol®), rosuvastatin (Crestor®), and simvastatin (Zocor®) (Figure 5-3).

Mechanism of action. The statins induce the partial inhibition of HMG-CoA reductase, the rate-limiting enzyme in cholesterol synthesis, thereby decreasing hepatic cholesterol production. The resultant decrease in intracellular cholesterol levels causes a compensatory upregulation of the LDL receptor, the membrane receptor that recognizes, binds, and internalizes circulating lipoproteins that carry apo B or apo E on their surfaces.[10] Thus, the administration of HMG-CoA reductase inhibitors results in the increased plasma clearance of cholesterol-rich LDL and, to a lesser extent, VLDL and intermediate-density lipoprotein (IDL). The hepatic synthesis of apo B-containing lipoproteins, such as VLDL, may also be affected by the statins, although the quantitative effect of this proposed mechanism is controversial.[11] The statins have different clinical properties, such as relative lipophilicity (eg, atorvastatin, simvastatin, and lovastatin are lipophilic, while pravastatin, fluvastatin, and rosuvastatin are more hydrophilic).[12]

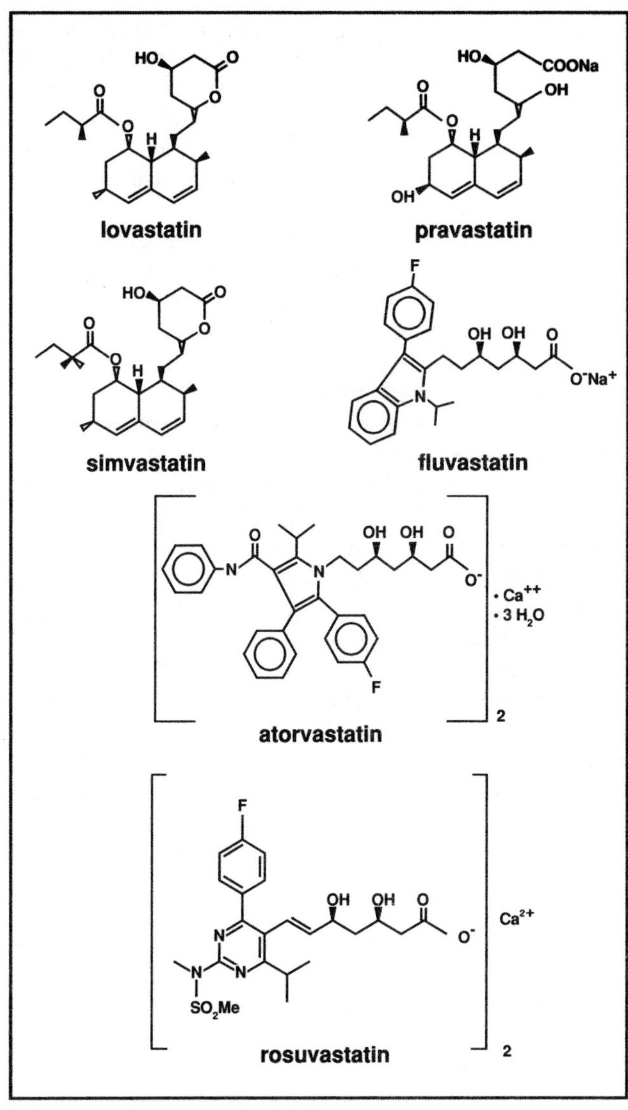

Figure 5-3: Structures of the HMG-CoA reductase inhibitors.

Statins also have pleiotropic effects that extend beyond LDL-C reduction.[13] These actions include reducing LDL oxidation, decreasing vascular expression of adhesion molecules, preserving endothelial synthesis of nitric oxide, and inhibiting thrombosis.[14] Moreover, statins may reduce coronary risk in the presence of systemic inflammation, even if LDL-C levels are not considered elevated.

Atherosclerosis is now regarded as fundamentally an inflammatory disease, and there is evidence to support the role of high-sensitivity C-reactive protein (hsCRP), a marker of systemic inflammation and a strong independent predictor of coronary events, in atherogenesis and plaque instability (see Appendix A, Case 6, for an example of the use of hsCRP in current clinical practice).[13] Moreover, hypothesis-generating data indicate that statins may reduce coronary risk in the presence of low LDL-C and high hsCRP levels.[13] The Justification for the Use of statins in Primary prevention: an Intervention Trial Evaluating Rosuvastatin (JUPITER) is being conducted to evaluate whether long-term statin therapy can reduce the incidence of major cardiovascular events in individuals who are at increased vascular risk as determined by an elevated hsCRP level, despite relatively low LDL-C levels (<130 mg/dL).[13] Support for this hypothesis in subjects without a history of CHD would strengthen the concept of treating high risk, rather than just high cholesterol, further blurring the distinction between primary and secondary prevention.

Dosage and efficacy. Both the established dosing ranges and the expected maximum LDL-C reduction with HMG-CoA reductase inhibitors are variable.

The dosing of lovastatin is 10 to 80 mg/d in a single dose or two divided doses. Lovastatin can achieve mean LDL-C reductions of up to 42% at the maximum dose. Simvastatin, which is dosed at 5 to 80 mg/d at bedtime, can achieve up to a 47% reduction in mean LDL-C levels. Pravastatin is dosed at 10 to 80 mg/d and results in a reduction of up to 37% in circulating LDL-C levels in patients with primary hypercholesterolemia.

Fluvastatin is dosed at 20 to 80 mg/d. The 80-mg tablet is available in an extended-release form only; however, 80 mg of immediate-release fluvastatin can be administered in two 40-mg doses. This agent reduces LDL-C by 20% to 35%. Atorvastatin is dosed at 10 to 80 mg/d and achieves LDL-C reductions of 40% to 60%. Rosuvastatin, which is dosed at 5 to 40 mg/d, produces LDL-C reductions of 45% to 63%. The statins are generally associated with moderate but predictable increases in HDL-C levels that may range from 5% to 15%. VLDL-C, which contains both apo B and apo E on its surface, may be cleared by the hepatocyte or peripheral tissues that have undergone upregulation of the LDL receptor, and levels of circulating TG may fall 10% to 37%. The statins do not appear to alter Lp(a) concentration significantly.

Statins are approved for use as an adjunct to diet, and the ATP III recommends a trial of TLC (including diet) before drug therapy is initiated. Recently, however, simvastatin was approved as an adjunct to diet from the start of therapy in high-risk patients. This is based on the results of the HPS (see Chapter 1).

Side effects and drug interactions: pooled data. The major abnormalities encountered clinically with these drugs are alterations of liver function and myopathy. However, the Prospective Pravastatin Pooling (PPP) Project, which is based on >112,000 person-years of experience, found no reported incidence of myopathy (ie, muscle ache or weakness together with creatine kinase [CK] levels >10x upper limit of normal [ULN]) with pravastatin or placebo. The incidence of serious hepatobiliary adverse events was similar in both treatment groups.[15] The PPP Project is discussed in Chapter 1.

Side effects and drug interactions: myopathy. Although the PPP Project provides ample data regarding the safety of statin therapy, physicians must be alert to the risks associated with any drug. A complete work-up at the first visit must include tests for hepatic, renal, and thyroid function (see Chapter 3, Figure 3-1). If initial treatment involving TLC does not achieve the desired effect, the patient's liver, kidney, and thyroid values should be reviewed before drug therapy is initiated. Baseline CK should also be tested. However, some patients receiving statin therapy may develop muscle toxicity below the threshold needed to raise serum CK levels.[16]

In one report, four cases of biopsy-confirmed myopathy were identified in the presence of normal CK levels.[16] The patients, who were subjects in a clinical trial, correctly identified blinded statin therapy on the basis of reproducible muscle symptoms. The pathologists who read the muscle biopsies were not blinded as to therapy. Statin levels were normal, but muscle biopsies showed evidence of mitochondrial dysfunction, and three patients had increased excretion of 3-methylglutaconic acid in the urine. The clinical significance of these findings is uncertain, and the frequency is unknown. However, CK level may not be an adequate test for statin-associated myopathy. Physicians must pay close attention to patients reporting muscle aches, decreased exercise tolerance, weakness, and other symptoms consistent with myopathy and determine whether the symptoms subside when statin therapy is withdrawn.

Clinically significant myopathy is uncommon with the statins; however, definite rhabdomyolysis occurs in approximately 0.1% of patients who receive statin monotherapy when rhabdomyolysis is defined as a 10-fold elevation of CK levels associated with a compatible symptom complex and myoglobinuria.[17] The mechanism by which the statins cause muscle toxicity has not been elucidated, but it may involve altered stability of membrane structure because of reduced cholesterol availability or decreased production of ubiquinone, which is used for electron transport by the mitochondria. The incidence of significant muscle toxicity is increased when the statins are combined with agents such as cyclosporine (Sandimmune®, Neoral®), gemfibrozil (Lopid®), erythromycin, nicotinic acid (Niacor®, Niaspan® Extended Release),[18] the protease inhibitors, nefazodone (Serzone®), and the macrolide antibiotics. The incidence of significant muscle toxicity with combination therapy may also be enhanced in patients with baseline renal insufficiency. Myopathy may also be more likely at higher statin doses (see discussion of A-to-Z trial, Chapter 1); in older patients (particularly thin or frail women); in the presence of multisystem disease, diabetes combined with chronic renal failure, or hypothyroidism; and following major surgery.[19]

Although rhabdomyolysis is a risk with all statins, the manufacturer of cerivastatin voluntarily withdrew the drug from the US and other markets in August 2001 because of a disproportionate number of rhabdomyolysis-associated deaths, especially when cerivastatin was combined with gemfibrozil. The withdrawal of cerivastatin has returned attention to the safety of lipid-modifying drugs. It is worth noting that the major clinical trials of statin therapy (see Chapter 1) have not reported an excess risk, and that the PPP Project found no difference between pravastatin and placebo in the incidence of serious noncardiovascular events, including deaths.[15] While the incidence of muscle symptoms may be increased when statin users are compared with nonusers, the absolute rates of myopathy, including rhabdomyolysis, are low, as reported in a large population-based survey.[20] In evaluating the risk-benefit ratio, physicians should weigh these concerns against the demonstrated benefits on clinical cardiovascular end points seen in a broad range of patients. As with all drugs, physicians should educate their patients about the potential side effects and benefits of statin treatment. Physical symptoms that have traditionally forewarned of muscle toxicity include muscle pain and brown urine.[17]

Side effects and drug interactions: liver function. All of the statins appear to be associated with some degree of transaminase elevation, although the incidence is generally less than 2% when toxicity is defined as an elevation of liver enzymes in excess of three times the ULN.[10-28] Liver function abnormalities appear to be most common in the first 4 to 12 months of therapy or after dose titration. For the first 3 months after treatment is begun or the dose is increased, liver function should be monitored at two 6-week intervals, followed by periodic monitoring every 4 to 6 months for the duration of therapy.

Side effects and drug interactions: renal. The prescribing information for rosuvastatin advises consideration of a dose reduction if patients on 40 mg experience unexplained persistent proteinuria during routine urinalysis.[29] This is based on an analysis of data from the rosuvastatin clinical trial program in which proteinuria (increase in urine dipstick levels from none, trace, or 1+ at baseline to ≥2+ at any visit) was generally transient and not associated with worsening renal function.[29] More recently, the data were reanalyzed according to revised criteria, including a redefinition of proteinuria (an increase from none or trace at baseline to ≥2+ at the last visit). Baseline levels of 1+ were excluded because they might indicate underlying renal dysfunction, which could produce an increase in urinary protein unrelated to treatment. In the reanalysis, the investigators describe the frequency of proteinuria as being generally similar with rosuvastatin (0.2% to 1.2%), comparator statins (0% to 1.1%), and placebo (0.6%).[30] Because the pooled reanalysis has several limitations, however, these results will require confirmation in well-designed clinical trials with prespecified renal end points. The proteinuria associated with rosuvastatin has been described as usually tubular (ie, characterized by primarily low molecular weight proteins).

A possible renoprotective effect of statins has also been proposed.[30] This is based on the observation that rosuvastatin-treated subjects in both short-term controlled trials (median duration of 8 weeks) and an open-label extension study (≥96 weeks) experienced no decrease in glomerular filtration rate (GFR), as estimated by the Modification of Diet in Renal Disease (MDRD) formula, even in the presence of preexisting renal disease (ie, baseline GFR <60 mL/min/1.73 m^2). These results are of interest and require further investigation and documentation as well.

Drug interactions: mechanisms. Investigators have recently become more interested in relationships among the molecular pathways of drug metabolism because they contribute to the risk for drug-drug interactions. Most statins are cleared via the cytochrome P-450 (CYP-450) pathway; thus, there is a possibility for drug interactions with competing substrates, inhibitors, or inducers of isoenzymes in this family.[31,32] Several CYP enzymes are involved in the biotransformation of fluvastatin, although in vitro studies indicate that it is metabolized primarily by CYP-2C9, a finding that has been confirmed in both blinded and nonblinded clinical trials.[33] Other HMG-CoA reductase inhibitors (ie, lovastatin, simvastatin, atorvastatin) are metabolized mainly by CYP-3A4.[17] Pravastatin and rosuvastatin are not extensively metabolized by the CYP-450 enzyme family, although rosuvastatin appears to undergo minimal metabolism by CYP-2C9.[34] Thus, pravastatin and rosuvastatin are less likely to produce adverse effects (eg, myopathy) when coadministered with drugs that inhibit CYP enzymes.

Side effects and drug interactions: combination therapy. The statins may be used in combination with other lipid-lowering agents whose differing mechanisms act in synergy with HMG-CoA reductase inhibition to enhance LDL-C response or modify other lipid subfractions. For example, resins and ezetimibe, a cholesterol absorption inhibitor, stimulate a compensatory increase in HMG-CoA reductase activity, albeit via different mechanisms of action. As a result, cholesterol may return to baseline levels. Adjunctive statin therapy can inhibit this secondary upregulation of the cholesterol biosynthetic pathway, thereby leading to an increase in LDL-receptor activity.

Patients may better tolerate combination therapy with a resin and a statin when lower doses of both drugs are administered. Statins have been combined with fibric acid derivatives (fibrates) in patients with mixed dyslipidemia. Fibrate therapy in a patient with hypertriglyceridemia can elevate LDL-C levels if there is an associated apo B/E (LDL) receptor defect, resulting in reduced LDL clearance. Although the combination of a fibric acid derivative and a statin theoreti-

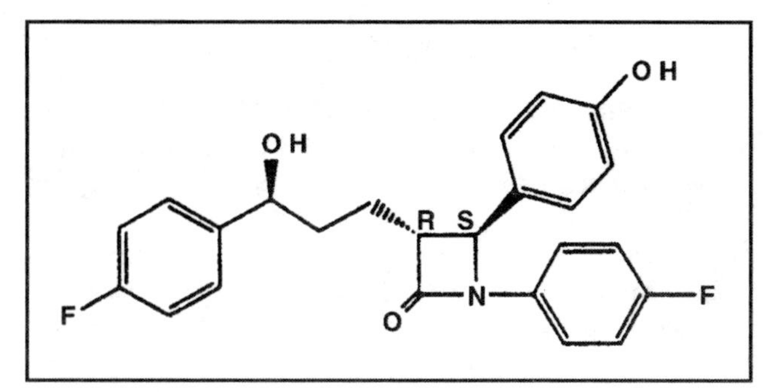

Figure 5-4: Chemical structure of ezetimibe.

cally would improve the efficacy of these agents, this combination is associated with an increased risk for myopathy (see Fibric Acid Derivatives, this chapter). Statin therapy, when prescribed in combination with nicotinic acid, also can increase the incidence of significant muscle toxicity. In 2001, the Food and Drug Administration approved a product combining extended-release nicotinic acid and lovastatin for the treatment of primary hypercholesterolemia and mixed dyslipidemia (Advicor®). When used appropriately, this agent can be a useful therapeutic option.

Cholesterol Absorption Inhibitors

Ezetimibe is the first in a new class of drugs known as cholesterol absorption inhibitors (Figure 5-4). As an adjunct to a standard cholesterol-lowering diet, ezetimibe is approved for the following: alone or in combination with a statin to reduce TC, LDL-C, and apo B in patients with primary (heterozygous familial and nonfamilial) hypercholesterolemia; in combination with atorvastatin, simvastatin, and/or other lipid-lowering treatments (eg, LDL apheresis) to reduce TC and LDL-C levels in patients with homozygous familial hypercholesterolemia (HoFH); and as monotherapy for the treatment of elevated sitosterol and campesterol levels in patients with homozygous familial sitosterolemia.[35]

The overwhelming evidence that reducing cholesterol levels decreases cardiovascular risk fuels the search for drugs with enhanced cholesterol-lowering activity and an improved side-effect profile, particularly in combination therapy. However, the precise effects of ezetimibe on cardiovascular morbidity and mortality have not yet been established.[35]

Mechanism of action. Cholesterol levels in the blood are regulated primarily by the liver, where cholesterol and bile acids are synthesized (endogenous pathway), and the intestine, which absorbs dietary and biliary cholesterol (exogenous pathway).[36] Cholesterol absorption inhibitors selectively limit the absorption of dietary and biliary cholesterol across the intestinal wall.[37,38] This decreases the delivery of intestinal cholesterol to the liver, reducing hepatic cholesterol stores and causing a compensatory increase in cholesterol clearance from the blood.[35] The absorption of other soluble food nutrients is not affected.[37] After oral administration, ezetimibe undergoes extensive first-pass glucuronidation in the intestinal wall. Both ezetimibe and its glucuronide conjugate undergo enterohepatic circulation, which repeatedly returns the drug to the intestine, its primary site of action.[37] Ezetimibe appears to localize at the brush border of the small intestine.[35]

Recently, researchers found that Niemann-Pick C1-like 1 (NPC1L1) protein is expressed in the brush border membrane of enterocytes and plays a critical role in intestinal cholesterol absorption. Moreover, when the effects of ezetimibe were tested in normal and NPC1L1-knockout mice, the latter were insensitive to the drug.[38] This suggests that NPC1L1 or an associated protein may be the molecular target of ezetimibe.

Dosage and efficacy. Ezetimibe is available in 10-mg tablets. The recommended dosage is 10 mg/d.[35] For the treatment of primary hypercholesterolemia, ezetimibe monotherapy has been shown to reduce TC, LDL-C, and apo B levels by 13%, 18%, and 14% to 16%, respectively.[35,39,40]

Because ezetimibe has a mechanism of action complementary to that of statins, it is suitable for use as adjunctive therapy. Clinical trials in hypercholesterolemic subjects at varying CHD risk levels have found that ezetimibe 10 mg plus a statin can lower TC, LDL-C, and apo B by up to approximately 43%, 59%, and 49% from baseline, respectively, depending on the statin, the statin dose, and the duration of treatment.[39,41,42]

The combination of ezetimibe 10 mg plus a statin at varying doses has been shown to decrease LDL-C levels more than the corresponding dose of statin alone; a similar effect on hsCRP levels was also found.[39] In addition, ezetimibe plus a statin at 10 or 20 mg can produce significantly greater reductions in LDL-C levels than statin monotherapy at doses up to 40 mg.[39] Combination therapy may also be more effective than a statin alone in helping high-risk hypercholesterolemic patients achieve an LDL-C goal <100 mg/dL.[42] Limited data are available concerning the effect of ezetimibe in non-Caucasian patients with primary hypercholesterolemia.[35]

Recently, an agent combining ezetimibe and simvastatin (Vytorin™) was introduced. In patients with primary hypercholesterolemia or mixed hyperlipidemia, ezetimibe/simvastatin is approved to reduce elevated TC, LDL-C, apo B, TG, and non-HDL-C and to increase HDL-C; in patients with HoFH, it is indicated for the reduction of TC and LDL-C as an adjunct to other lipid-lowering treatments (eg, LDL apheresis). Ezetimibe/simvastatin is available in 10/10-mg, 10/20-mg, 10/40-mg, and 10/80-mg doses.[43]

In patients with homozygous sitosterolemia, ezetimibe can significantly lower plasma sitosterol and campesterol by 21% and 24%, respectively, vs increases of 4% and 3% with placebo.[35]

Side effects and drug interactions. Although cholesterol absorption inhibitors localize in the intestine, they may be less likely than resins to cause gastrointestinal side effects.

As reported in the prescribing information, clinical trials of ezetimibe vs placebo found that adverse events occurred with similar frequency in both treatment groups. The incidence of consecutive elevations (≥3 x ULN) in serum transaminase levels was 0.5% with ezetimibe vs 0.3% with placebo.[35] Other adverse events with incidence rates of 2% to 4% in each group included fatigue, abdominal pain, diarrhea, arthralgia, back pain, coughing, and infection (eg, pharyngitis, sinusitis).[35] In trials comparing ezetimibe plus a statin vs statin therapy alone, consecutive serum transaminase elevations were more frequent with combination therapy: 1.3% vs 0.4%. The precautions for ezetimibe specify that liver function tests must be performed before initiating combination therapy and thereafter as recommended for each statin. Combination therapy is contraindicated in patients with active liver disease or with unexplained, persistent elevations in serum transaminase levels.[35] Adverse events occurring with ezetimibe/simvastatin in ≥2% of clinical trial participants include headache, influenza, upper respiratory tract infection, myalgias, and pain in an extremity.[43]

Because ezetimibe is a recently approved drug, available safety and tolerability data are limited. Additional data on long-term use in diverse populations are needed, both in clinical trials and in a variety of patient-care settings.

Drug interactions: mechanisms. Ezetimibe has shown no significant interaction with cytochrome P-450 (CYP-450) enzymes, which are largely responsible for the metabolism of statins (excluding pravastatin and perhaps rosuvastatin).[34,37] Consequently, the potential for drug interactions with substrates of this enzyme system appears to be low. In the case of ezetimibe/simvastatin, however, potent inhibitors of CYP3A4 can increase the risk for myopathy by interfering with the elimination of the simvastatin component. These inhibitors include azole antifungals, macrolide antibiotics, protease inhibitors, nefazodone, cyclosporine, and grapefruit juice (>1 quart daily). Caution is also needed when prescribing ezetimibe/simvastatin with other lipid-lowering drugs (eg, gemfibrozil, nicotinic acid) that can cause myopathy when administered alone.[43] In addition, fenofibrate (TriCor®) and gemfibrozil can raise ezetimibe concentrations by 1.5-fold and 1.7-fold, respectively.[43]

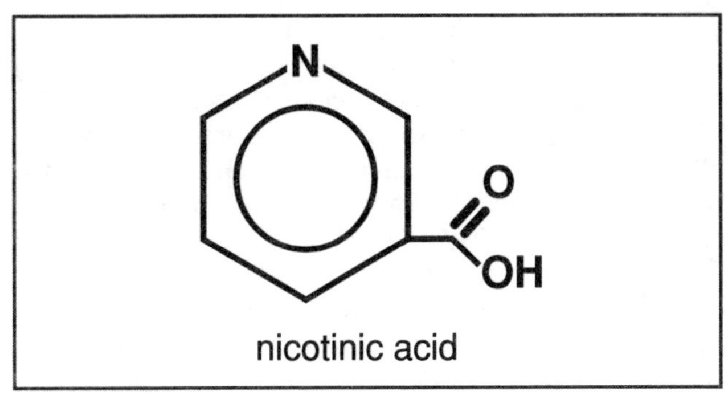

Figure 5-5: Structure of nicotinic acid (niacin).

Treating Mixed Dyslipidemias

Nicotinic Acid

Nicotinic acid (also called niacin, but not to be confused with nicotinamide) is an essential B vitamin that was determined to improve the lipid profile at dosage levels that far exceed the amounts necessary to prevent deficiency. Nicotinic acid (Figure 5-5) exhibits beneficial effects on multiple lipoprotein subclasses and has been extensively tested in primary- and secondary-prevention trials using both combination and monotherapy regimens.[44]

Mechanism of action. Nicotinic acid favorably modifies the lipid profile and is effective in all dyslipidemias with the exception of Fredrickson type I hyperlipidemia, which is characterized by increased levels of chylomicrons (see Chapter 3, Table 3-9). Nicotinic acid exerts a primary hepatic effect that decreases the production and release of VLDL. The formation of VLDL is the first step in the endogenous lipid cascade, and reduced synthesis or release of the initial lipoprotein results in decreased levels of all subsequent particles, including VLDL remnants and LDL. In addition to the primary reduction in hepatic release of VLDL, nicotinic acid may affect VLDL clearance from the plasma compartment. Nicotinic acid also has the peripheral effect of reducing the release of free fatty acids from the adipocyte, resulting in diminished availability of these metabolic precursors for TG production. The peripheral activity of nicotinic acid may be transient and not quantitatively significant with chronic therapy. Probably the most effective HDL-C-raising agent, nicotinic acid appears to exert its beneficial effect on HDL in part by decreasing the fractional catabolic rate of apo A-I.[45] Unlike other conventional lipid-modifying therapies (ie, statins, resins, and fibrates), nicotinic acid significantly lowers Lp(a); however, the mechanism is not completely delineated. Nicotinic acid also has been associated with reductions in the LDL subclass Pattern B, which is characterized by atherogenic small, dense LDL particles.

Dosage and efficacy. Nicotinic acid administered as a vitamin to prevent deficiency requires a dosing level of 1 to 5 mg/d. However, significantly higher doses are required to exert a lipid-modifying effect. One gram of nicotinic acid per day may increase HDL-C concentrations, but dosing ranges of 2 to 6 g/d are necessary to attain the maximum expected benefit of nicotinic acid on other lipoprotein subfractions. Low-density lipoprotein cholesterol may be expected to decline by 5% to 25%, and TG levels may fall 20% to 50%. Depending on pretreatment TG levels, the response of HDL-C to nicotinic acid therapy varies and may range from 15% to 35%. Nicotinic acid therapy also may be beneficial in the treatment of familial defective apo B-100, a dyslipidemia in which the circulating levels of LDL-C are increased because of impaired receptor removal caused by an abnormality in the structure of apo B-100, which is the receptor ligand. By decreasing the precursors of LDL, nicotinic acid may alter the levels of this particle without requiring increased receptor-mediated clearance.

Side effects and drug interactions: flushing. The use of nicotinic acid has been limited because of a variety of side effects that frequently result in decreased patient compliance.[46] Most side effects associated with nicotinic acid, while troublesome, may be self-limiting or minimized with effective patient counseling and dosing regimens. The most common side effect is cutaneous flushing, which is prostaglandin-mediated and may be severe enough to cause hypotension. The flushing is frequently accompanied by severe pruritus and may be minimized by beginning at low doses (50 mg t.i.d.) and increasing the dose at intervals of 4 to 7 days (standard formulation) or every 4 weeks (sustained-release formulation). Pretreatment with aspirin may blunt the prostaglandin-mediated vasodilatation and should be considered a necessary concomitant medication if there are no contraindications.

Side effects and drug interactions: liver function. The other effects of nicotinic acid are considerably more troublesome and include potentially fatal complications, such as fulminant hepatic failure. Abnormally elevated circulating transaminase levels are seen in approximately 5% of patients whose dosing ranges of nicotinic acid exceed 3 g/d. Because the transaminase elevations are generally asymptomatic, periodic monitoring of liver function is required. A sudden, unexplained fall in circulating cholesterol levels may herald impending severe hepatic dysfunction; this complication may be more common when switching from crystalline nicotinic acid to the sustained-release form. Mild, asymptomatic, and persistent elevations of liver enzymes are not an absolute indication for termination of nicotinic acid therapy, and dose reductions of up to 50% may be beneficial in an attempt to determine if the liver function will return to normal. However, persistent transaminase elevations in excess of three times the normal levels should result in discontinuation of the drug.

Side effects and drug interactions: metabolic. A variety of adverse metabolic effects may also be associated with nicotinic acid therapy, including worsening of glycemic control (which may occur in up to 10% of patients). The precise mechanism involved in the alteration of glucose tolerance or induction of diabetes after nicotinic acid therapy is complicated and may be secondary to alteration of peripheral sensitivity to insulin. Diabetes is not an absolute contraindication to the use of nicotinic acid therapy, but this agent should be used with caution in individuals with metabolic abnormalities indicating possible insulin resistance (eg, type 2 diabetes, fasting plasma glucose ≥100 mg/dL, other risk factors for the metabolic syndrome).

Side effects and drug interactions: myopathy. Although myopathy is uncommon with nicotinic acid monotherapy, the risk is increased when it is combined with other agents such as the HMG-CoA reductase inhibitors or fibric acid derivatives. To minimize the possibility of myopathy, baseline CK should be measured before prescribing nicotinic acid as monotherapy or as combination therapy with a statin or fibrate drug. If patients taking nicotinic acid experience symptoms compatible with myopathy, therapy should be stopped immediately and CK levels retested. See HMG-CoA Reductase Inhibitors for more information on statin-associated myopathy in the presence of normal CK levels.[16]

Side effects and drug interactions: other. Hyperuricemia and precipitation of gouty arthritis have been reported with nicotinic acid therapy. Uric acid elevations may occur in 5% to 10% of patients receiving niacin; patients who are predisposed to gout should be informed of this potential complication.

Gastrointestinal symptoms are common with nicotinic acid monotherapy, and exacerbation of peptic ulcer disease may be encountered. Nonspecific symptoms, such as nausea or mild abdominal discomfort, may occur in up to 20% of patients receiving nicotinic acid therapy. These may be further exacerbated by the concomitant administration of aspirin in an attempt to minimize the cutaneous complications.

Fibric Acid Derivatives

A variety of fibric acid derivatives, or fibrates, are used worldwide.[47] In the United States, however, gemfibrozil and fenofibrate are the available agents of this class (Figure 5-6).

Mechanism of action. The fibric acid derivatives have a complex mechanism of action that produces multiple effects on both the synthetic and catabolic pathways of TG-rich particles.[48]

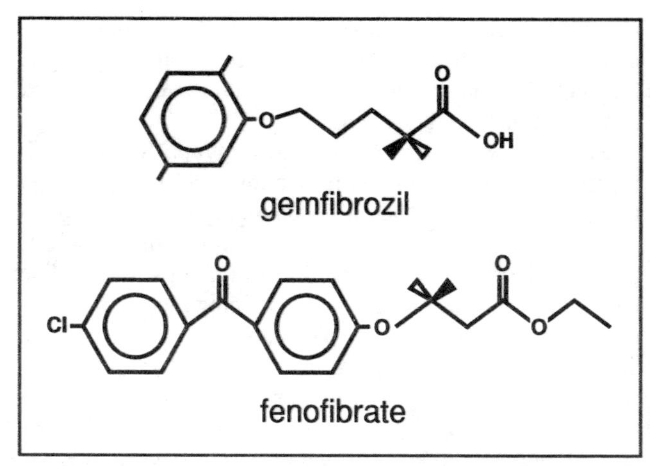

Figure 5-6: Structures of the fibrates available in the United States.

The fibrates trigger peroxisome proliferator-activated receptors (PPARs), a subfamily of nuclear receptors that controls a variety of cellular functions, such as lipid and lipoprotein metabolism, fatty acid oxidation, glucose metabolism, adipogenesis, and cellular differentiation. Activated PPARs may regulate the transcription of several genes, including those that can increase expression of apo A-I or adenosine triphosphate binding cassette (ABC) A1, a transporter protein involved in apo A-I-mediated cholesterol efflux. In addition, PPARs may have direct vascular effects that influence atherosclerosis at the level of the vessel wall.

Fibric acid derivatives increase the catabolism of TG-rich lipoproteins by activating the enzyme lipoprotein lipase, which catalyzes the hydrolysis of chylomicrons and VLDL, resulting in the formation of remnant particles and IDL, respectively. The activation of lipoprotein lipase also results in the transfer of apolipoproteins and surface phospholipids from VLDL to HDL. This plays a role in the frequently observed inverse relation between TG and HDL-C. The effect of fibric acid derivatives on LDL is complex and is a function of the relative activity of the LDL receptors and pretreatment TG levels. Activation of lipoprotein lipase in a hypertriglyceridemic patient with defective or reduced activity of the apo B/E receptor may increase the catabolism of VLDL, resulting in an increase in LDL particles that cannot be effectively cleared from the circulation. Hypertriglyceridemic patients frequently have increased concentrations of atherogenic small, dense LDL particles. Fibric acid derivative therapy may result in the remodeling of LDL into a larger, buoyant, and potentially less atherogenic form.[49]

Dosage and efficacy. Dosage information on fibrates can be found in Table 5-6. Fibrates are predominantly used in patients with elevated TG or low HDL-C to lower circulating levels of TG, and they may decrease TG levels by 20% to 50%, with an accompanying increase in HDL-C of 5% to 20%. After the administration of fibric acid derivatives, LDL-C levels are unpredictable, but may be expected to decrease by 10% to 15%. Fibric acid derivatives may have effects beyond altering the circulating lipid concentrations, including reductions in fibrinogen, factor VII phospholipid complex, and plasminogen activator inhibitor-1 (PAI-1). Fibric acid derivatives may also have beneficial effects on epinephrine-induced platelet aggregation. No substantial effect on Lp(a) has been consistently demonstrated with the fibrates.

Side effects and drug interactions. The chief side effects of fibric acid derivatives are gastrointestinal and may occur in up to 5% of patients receiving these agents. Earlier intervention trials implied an increase in noncardiac morbidity and mortality after the use of clofibrate, a drug no longer available in the United States, although safety data regarding other agents of this class have documented no definite adverse effects. Liver function abnormalities may occasionally be noted, but hepatic failure or progression to chronic liver disease has not been documented with these compounds. Myopathy is uncommon with the use of fibric acid derivatives as monotherapy, but it may increase when these agents are combined with drugs such as HMG-CoA reductase inhibitors. The adverse interaction between fibrates and statins, with the induction of myopathy, is especially accentuated in subjects with renal insufficiency. In August 2001, cerivastatin was withdrawn from the US market because of the incidence of rhabdomyolysis. Between 1998, when cerivastatin was introduced, and December 2000, 8 months before it was withdrawn, slightly more than 50% of the rhabdomyolysis cases were associated with concomitant fibrate use.[50]

The combined use of statins and fibrates should be avoided unless the benefit of further lipid modification is likely to outweigh the increased risk of this drug combination. To minimize the possibility of adverse effects, particularly myopathy, a fibrate should not be added to higher-dose statin therapy; gemfibrozil should be avoided (fenofibrate may be considered, unless contraindicated); combination therapy should not be used in patients with renal insufficiency (fibrates are renally excreted), those taking drugs that interfere with statin clearance, or patients >70 years of age; and pretreatment and periodic testing of CK levels is necessary in all cases. Physicians must tell their patients to report muscle pain, weakness, severe malaise, or urine discoloration, and to stop therapy immediately if these symptoms occur. If the symptoms are accompanied by elevated CK levels, the physician may wait until CK values return to normal and then initiate a second trial with low doses of different drugs while closely monitoring the patient for symptoms and elevated CK levels. Recently, patients in a clinical trial were found to have statin-associated myopathy in the presence of normal CK levels[16] (see HMG-CoA Reductase Inhibitors). Because fibrates potentiate the action of warfarin, prothrombin time should be monitored when these drugs are administered to patients taking anticoagulant therapy.

Combination Agents

The cost of medications to treat chronic conditions, particularly in the elderly, is growing.[51] Combination products may help reduce drug costs and facilitate compliance with multidrug regimens. In addition to ezetimibe/simvastatin and nicotinic acid/lovastatin, there are two other agents that combine an HMG-CoA reductase inhibitor with a second drug.

Buffered aspirin/pravastatin (Pravigard™) is approved for use in patients with clinical evidence of CVD to reduce total mortality by decreasing coronary death and to reduce the risk for MI, myocardial revascularization procedures, and stroke/transient ischemic attack. It is also indicated to slow the progression of coronary atherosclerosis.[52]

Amlodipine/atorvastatin (Caduet®) combines a long-acting calcium channel blocker to treat hypertension or chronic stable angina with the HMG-CoA reductase inhibitor atorvastatin. This product is indicated in patients for whom treatment with both amlodipine and atorvastatin is appropriate.[53]

As with all medications, the complete prescribing information for combination agents should be consulted.

Over-the-Counter Statins

Recently, the first over-the-counter (OTC) statin, simvastatin, was launched in Britain. The availability of low-dose statin therapy without a prescription raises several questions, including the following: Will this lead to an increase in the number of eligible patients receiving treatment? Will consumers be able to self-administer the drugs appropriately and safely? What will happen in the case of persons who need more aggressive treatment or clinical monitoring?[54] The anticipated publication of a large clinical use study of OTC statin therapy may help answer these and other questions.

References

1. Executive Summary of the Third Report of the National Cholesterol Education Program (NCEP) Expert Panel on Detection, Evaluation, and Treatment of High Blood Cholesterol in Adults (Adult Treatment Panel III). *JAMA* 2001;285:2486-2497.

2. Grundy SM, Cleeman JI, Merz CNB, et al: Implications of recent clinical trials for the National Cholesterol Education Program Adult Treatment Panel III guidelines. *Circulation* 2004; 110:227-239.

3. Krauss RM, Eckel RH, Howard B, et al: AHA Dietary Guidelines, Revision 2000: a statement for healthcare professionals from the Nutrition Committee of the American Heart Association. *Circulation* 2000;102:2296-2311.

4. Cannon CP, Braunwald E, McCabe CH, et al: Intensive versus moderate lipid lowering with statins after acute coronary syndromes. *N Engl J Med* 2004;350:1495-1504.

5. Lloyd-Jones DM, Nam BH, D'Agostino RB Sr, et al: Parental cardiovascular disease as a risk factor for cardiovascular disease in middle-aged adults: a prospective study of parents and offspring. *JAMA* 2004;291:2204-2211.

6. Follow-up report on the diagnosis of diabetes mellitus: The Expert Committee on the Diagnosis and Classification of Diabetes Mellitus. *Diabetes Care* 2003;26:3160-3167.

7. Grundy SM, Brewer HB, Cleeman JI, et al, for the Conference participants: Definition of metabolic syndrome: Report of the National Heart, Lung, and Blood Institute/American Heart Association Conference on Scientific Issues Related to Definition. *Circulation* 2004;109:433-438.

8. Farmer JA, Gotto AM Jr: Currently available hypolipidaemic drugs and future therapeutic developments. *Baillieres Clin Endocrinol Metab* 1995;9:825-847.

9. Schectman G, Hiatt J: Dose-response characteristics of cholesterol-lowering drug therapies: implications for treatment. *Ann Intern Med* 1996;125:990-1000.

10. Hsu I, Spinler SA, Johnson NE: Comparative evaluation of the safety and efficacy of HMG-CoA reductase inhibitor monotherapy in the treatment of primary hypercholesterolemia. *Ann Pharmacother* 1995;29:743-759.

11. Burnett JR, Wilcox LJ, Telford DE, et al: Inhibition of HMG-CoA reductase by atorvastatin decreases both VLDL and LDL apolipoprotein B production in miniature pigs. *Arterioscler Thromb Vasc Biol* 1997;17:2589-2600.

12. Hamelin BA, Turgeon J: Hydrophilicity/lipophilicity: relevance for the pharmacology and clinical effects of HMG-CoA reductase inhibitors. *Trends Pharmacol Sci* 1998;19:26-37.

13. Ridker PM, JUPITER Study Group: Rosuvastatin in the primary prevention of cardiovascular disease among patients with low levels of low-density lipoprotein cholesterol and elevated high-

sensitivity C-reactive protein: rationale and design of the JUPITER Trial. *Circulation* 2003;108:2292-2297.

14. Vaughan CJ, Gotto AM Jr, Basson CT: The evolving role of statins in the management of atherosclerosis. *J Am Coll Cardiol* 2000;35:1-10.

15. Pfeffer MA, Keech A, Sacks FM, et al: Safety and tolerability of pravastatin in long-term clinical trials. Prospective Pravastatin Pooling (PPP) Project. *Circulation* 2002;105:2341-2346.

16. Phillips PS, Haas RH, Bannykh S, et al: Statin-associated myopathy with normal creatine kinase levels. *Ann Intern Med* 2002; 137:581-585.

17. Farmer JA, Torre-Amione G: Comparative tolerability of the HMG-CoA reductase inhibitors. *Drug Saf* 2000;23:197-213.

18. Garnett WR: Interactions with hydroxymethylglutaryl-coenzyme A reductase inhibitors. *Am J Health Syst Pharm* 1995;52: 1639-1645.

19. Pasternak RC, Smith SC Jr, Bairey-Merz CN, et al: ACC/AHA/NHLBI clinical advisory on the use and safety of statins. *J Am Coll Cardiol* 2002;40:567-572.

20. Gaist D, Rodriguez LA, Huerta C: Lipid-lowering drugs and risk of myopathy: a population-based follow-up study. *Epidemiology* 2001;12:565-569.

21. Boccuzzi SJ, Bocanegra TS, Walker JF, et al: Long-term safety and efficacy profile of simvastatin. *Am J Cardiol* 1991;68:1127-1131.

22. Bradford RH, Shear CL, Chremos AN, et al: Expanded Clinical Evaluation of Lovastatin (EXCEL) study results. I. Efficacy in modifying plasma lipoproteins and adverse event profiles in 8,245 patients with moderate hypercholesterolemia. *Arch Intern Med* 1991;151:43-49.

23. Bradford RH, Downtown M, Chremos AN, et al: Efficacy and tolerability of lovastatin in 3,390 women with moderate hypercholesterolemia. *Ann Intern Med* 1993;118:850-855.

24. Bradford RH, Shear CL, Chremos AN, et al: Expanded Clinical Evaluation of Lovastatin (EXCEL) study results: two-year efficacy and safety follow-up. *Am J Cardiol* 1994;17:667-673.

25. Marais AD, Firth JC, Bateman ME, et al: Atorvastatin: an effective lipid-modifying agent in familial hypercholesterolemia. *Arterioscler Thromb Vasc Biol* 1997;17:1527-1531.

26. Stein EA, Davidson MH, Dobs AS, et al: Efficacy and safety of simvastatin 80 mg/day in hypercholesterolemic patients. The Expanded Dose Simvastatin US Study Group. *Am J Cardiol* 1998; 82:311-316.

27. Marz W, Wollschlager H, Klein G, et al: Safety of low-density lipoprotein cholesterol reduction with atorvastatin versus simvastatin in a coronary heart disease population (the TARGET TANGIBLE trial). *Am J Cardiol* 1999;84:7-13.

28. Ose L, Davidson MH, Stein EA, et al: Lipid-altering efficacy and safety of simvastatin 80 mg/day: long-term experience in a large group of patients with hypercholesterolemia. World-Wide Expanded Dose Simvastatin Study Group. *Clin Cardiol* 2000;23: 39-46.

29. Crestor® (rosuvastatin calcium) Tablets. Prescribing information. ©AstraZeneca 2003. Revised 08/03.

30. Vidt DG, Cressman MD, Harris S, et al: Rosuvastatin-induced arrest in progression of renal disease. *Cardiology* 2004;102:52-60.

31. Bottorf MB: Distinct drug-interaction profiles for statins. *Am J Health Syst Pharm* 1999;56:1019-1020.

32. Beaird SL: HMG-CoA reductase inhibitors: assessing differences in drug interactions and safety profiles. *J Am Pharm Assoc (Wash)* 2000;40:637-644.

33. Scripture CD, Pieper JA: Clinical pharmacokinetics of fluvastatin. *Clin Pharmacokinet* 2001;40:263-281.

34. Olsson AG, McTaggart F, Raza A: Rosuvastatin: a highly effective new HMG-CoA reductase inhibitor. *Cardiovasc Drug Rev* 2002;20:303-328.

35. Zetia™ (ezetimibe) tablets. Merck/Schering-Plough Pharmaceuticals. © Merck/Schering-Plough Pharmaceuticals, 2001, 2002.

36. Kosoglou T, Meyer I, Veltri EP, et al: Pharmacodynamic interaction between the new selective cholesterol absorption inhibitor ezetimibe and simvastatin. *Br J Clin Pharmacol* 2002;54:309-319.

37. Bays HE, Moore BP, Drehobl MA, et al: Effectiveness and tolerability of ezetimibe in patients with primary hypercholesterolemia: pooled analysis of two phase II studies. *Clin Ther* 2001; 23:1209-1230.

38. Altmann SW, Davis HR Jr, Zhu LJ, et al: Niemann-Pick C1 Like 1 protein is critical for intestinal cholesterol absorption. *Science* 2004;303:1201-1204.

39. Ballantyne CM, Houri J, Notarbartolo A, et al: Effect of ezetimibe coadministered with atorvastatin in 628 patients with primary hypercholesterolemia. A prospective, randomized, double-blind trial. *Circulation* 2003;107:2409-2415.

40. Davidson MH, McGarry T, Bettis R, et al: Ezetimibe coadministered with simvastatin in patients with primary hypercholesterolemia. *J Am Coll Cardiol* 2002;40:2125-2134i.

41. Ballantyne CM, Blazing MA, King TR, et al: Efficacy and safety of ezetimibe co-administered with simvastatin compared with atorvastatin in adults with hypercholesterolemia. *Am J Cardiol* 2004;93:1487-1494.

42. Feldman T, Koren M, Insull W Jr, et al: Treatment of high-risk patients with ezetimibe plus simvastatin co-administration versus simvastatin alone to attain National Cholesterol Education Program Adult Treatment Panel III low-density lipoprotein cholesterol goals. *Am J Cardiol* 2004;93:1481-1486.

43. Vytorin™ 10/10 (ezetimibe 10 mg/simvastatin 10 mg tablets), Vytorin™ 10/20 (ezetimibe 10 mg/simvastatin 20 mg tablets), Vytorin™ 10/40 (ezetimibe 10 mg/simvastatin 40 mg tablets), Vytorin™ 10/80 (ezetimibe 10 mg/simvastatin 80 mg tablets). North Wales, PA, Merck/Schering-Plough Pharmaceuticals. Issued August 2004.

44. Brown WV: Niacin for lipid disorders. Indications, effectiveness, and safety. *Postgrad Med* 1995;98:185-193.

45. Jin FY, Kamanna VS, Kashyap ML: Niacin decreases removal of high-density apolipoprotein A-I but not cholesterol ester by Hep G2 cells. Implication for reverse cholesterol transport. *Arterioscler Thromb Vasc Biol* 1997;17:2020-2028.

46. Gibbons LW, Gonzalez V, Gordon N, et al: The prevalence of side effects with regular and sustained-release nicotinic acid. *Am J Med* 1995;99:378-385.

47. Miller DB, Spence JD: Clinical pharmacokinetics of fibric acid derivatives (fibrates). *Clin Pharmacokinet* 1998;34:155-162.

48. Duez H, Fruchart JC, Staels B: PPARs in inflammation, atherosclerosis and thrombosis. *J Cardiovasc Risk* 2001;8:187-184.

49. Chapman MJ, Guérin M, Bruckert E: Atherogenic, dense low-density lipoproteins. Pathophysiology and new therapeutic approaches. *Eur Heart J* 1998;19(suppl A):A24-A30.

50. Fischer C, Wolfe SM, Sasich L, et al: Petition to the FDA to issue strong warnings about the potential for certain cholesterol-lowering drugs to cause potentially life-threatening muscle damage [letter]. HRG Publication #1588. *Public Citizen.* The Health Research Group. August 20, 2001, 5 pp.

51. Fischer MA, Avorn J: Economic implications of evidence-based prescribing for hypertension. Can better care cost less? *JAMA* 2004;291:1850-1856.

52. Pravigard™ PAC (buffered aspirin and pravastatin sodium) tablets. Princeton, NJ, Bristol-Myers Squibb Company. Issued May 2003.

53. Caduet® (amlodipine besylate/atorvastatin calcium) tablets. Dublin, Ireland: Pfizer Ireland Pharmaceuticals, ©2004.

54. Gotto AM Jr: The case for over-the-counter statins [editorial]. *Am J Cardiol* 2004;94:753-756.

Special Populations

Historically, research on coronary heart disease (CHD) and on the effects of lipid-modifying drug therapy was conducted in middle-aged men, a high-risk group. Subgroup analyses helped guide the treatment of other populations until recently, when the results of clinical trials designed to study previously underrepresented patient groups became available.

In 2004, the Adult Treatment Panel III (ATP III) of the National Cholesterol Education Program (NCEP) published a report that revises certain aspects of its 2001 guidelines on detecting, evaluating, and treating high blood cholesterol in adults.[1,2] These revisions are based on recent evidence from large, randomized clinical trials studying the effects of cholesterol-lowering drug therapy on cardiovascular end points in a range of subjects, including the elderly, women, people with diabetes, and those at high risk despite relatively low cholesterol levels according to Western standards.[2] Also in 2004, the American Heart Association (AHA) issued evidence-based guidelines for cardiovascular disease prevention in women.[3]

This chapter reviews issues specific to these special populations and examines the pertinent new or revised guidelines.

Women

In 2001, the last year for which data are available, approximately 250,000 women died of CHD. This represents 49% of all CHD deaths.[4] Moreover, 64% of women who

Table 6-1: Special Considerations in Treating Dyslipidemia in Women

Menopausal Status	Treatment
Premenopausal Women	In some patients, higher LDL-C cut points for initiating drug therapy are appropriate because of related premenopausal cardioprotection. Lifestyle changes may suffice in many premenopausal women. If drug therapy is necessary, resins are particularly suitable for lowering LDL-C in women considering pregnancy because there is low systemic exposure. Although ezetimibe has minimal systemic absorption,[5] its effect on an unborn baby is unknown (FDA pregnancy category C). Therefore, its use in women who are pregnant or may become pregnant is not advised. Statins are contraindicated during pregnancy, and women of childbearing age on statin therapy should also use a contraceptive (see Appendix A, Case 5).

died suddenly of CHD had no prior symptoms. When considered with age, sex constitutes a major risk factor for coronary disease. Premenopausal women are at very low risk for CHD; in postmenopausal women, CHD rates are two to three

Menopausal Status	Treatment
Postmenopausal Women	Although HRT has been shown to lower LDL-C and raise HDL-C levels, it does not appear to slow atherosclerosis progression. Furthermore, recent evidence indicates that estrogen either alone or with progestin increases cardiovascular risk in the short term and suggests that this therapy may also increase risk over the longer treatment period needed to prevent chronic disease. Based on results of the Women's Health Initiative (this chapter), the FDA has approved the use of HRT at the lowest dose and for the shortest duration needed to relieve vasomotor symptoms. For other postmenopausal indications, nonestrogen products should be considered.*

*http://www.fda.gov/bbs/topics/NEWS/2003/NEW00863.html
HDL-C = high-density lipoprotein cholesterol; HRT = hormone replacement therapy; LDL-C = low-density lipoprotein cholesterol; FDA = Food and Drug Administration.

times higher than those in age-matched premenopausal women.[4] Recent evidence has indicated that the use of estrogen either alone or with progestin in postmenopausal women increases the risk for cardiovascular disease (Table 6-1).

The risk factors associated with CHD in women are largely the same as those affecting men; however, the relative weight of these risk factors may differ between the sexes. The reasons for these apparent differences are still unclear.[6]

Risk Factors for Premenopausal Women

Although premenopausal women have a much lower rate of atherosclerosis than men, some variables increase a young woman's chance for heart disease. Young female smokers 16 to 44 years of age are 10 times more likely to suffer myocardial infarction (MI) than nonsmokers.[7] Although long-term follow-up of oral contraceptive (OC) users does not suggest that they face increased risk for CHD, heavy smokers who use OCs have a 20- to 30-fold increased risk for MI compared with smokers who do not use OCs.[8,9] In addition, women with lupus erythematosus, type 1 or type 2 diabetes, or artificial menopause all have a greater chance of developing CHD compared with women in the general population.[10-14] The presence of diabetes increases the risk for CHD in women, thereby markedly attenuating the relative protection experienced before menopause (see next section).[11,12] Women with polycystic ovary syndrome have an increase in the number of cardiovascular risk factors (eg, diabetes), although their rates of morbidity and mortality from CHD are not as high as previously predicted.[15,16]

Risk Factors for Postmenopausal Women

Under 20 years of age, men and women have similar lipid levels, but thereafter, their lipid profiles tend to diverge. Between the ages of 20 and 50, total cholesterol (TC) levels are generally higher in men than in women. After age 55, however, the cholesterol levels of women increase and may slightly exceed those of men. Furthermore, triglyceride (TG) levels generally increase in women during middle age, whereas they may decrease in men. Women also experience a decrease in high-density lipoprotein cholesterol (HDL-C) levels following menopause (although their HDL-C generally remains about 10 mg/dL higher than the level in men).[17]

To examine the prevalence and degree of carotid atherosclerosis in women before and after menopause, researchers compared data from 292 premenopausal and 294 postmenopausal women enrolled in the Women's Healthy Lifestyle Project (WHLP) and the Healthy Women Study (HWS), respectively.[18] In both studies, mean baseline age was 47 years, although women in the HWS were an average of 4 months older, a statistically significant difference. Subclinical carotid atherosclerosis was observed in both cohorts, but disease prevalence was 4.1 and 5.3 times greater in women at 5 and 8 years after menopause, respectively, compared with premenopausal women. In addition, subclinical disease at 5 and 8 years after menopause was significantly associated with premenopausal risk factors, including smoking and low-density lipoprotein cholesterol (LDL-C) level. Together with the fact that CVD is the leading cause of death in women, these results highlight both the progressive nature of atherosclerotic vascular disease and the importance of risk factor modification in younger women in order to prevent the development of clinically significant atherosclerosis in later years.[18]

In the HWS,[18] carotid artery wall thickness was significantly related to increases in LDL-C, TG, insulin, body mass index (BMI), and systolic blood pressure. There was also a significant inverse relation between arterial wall thickening and HDL-C levels.

Although excess weight affects women's cholesterol levels, weight loss does not appear to improve the lipid profile unless it is accompanied by physical exercise.[19] Furthermore, body weight distribution is an important risk factor in women.[20] For example, weight gain around the waist is associated with lower HDL-C levels,[19,20] and abdominal fat may confer a particularly high CHD risk.[21-23]

While diabetes increases CHD risk in both diabetic vs nondiabetic women and diabetic vs nondiabetic men, the differential is markedly greater in women than in men, abolishing the female vs male advantage.[6,24] However, the role of age in the relation between diabetes and CVD in women is potentially that of a confounding variable because of its complex association with both an increasing prevalence of diabetes and a loss in life expectancy because of CVD.[24] Nevertheless, a 65-year-old woman in the United States can expect to live for almost 2 more decades (19.2 years), and approximately one third of the US population lives to more than 85 years of age.[24] For many individuals, these years will be lived with a history of CHD and diabetes. Therefore, aggressive attention to modifiable CHD risk factors is important in order to retard disease progression and maintain or improve quality of life in older women and men.[24]

American Heart Association Guidelines

The 2004 AHA guidelines for the prevention of CVD in women[3] were developed by a group of experts representing several public and private educational and advocacy organizations, including the American College of Cardiology, the American College of Obstetricians and Gynecologists, the Centers for Disease Control and Prevention, and the National Heart, Lung and Blood Institute. Organizations that endorse the guidelines include the American Diabetes Association (ADA) and the American Geriatrics Society.

The authors conducted a systematic search of the literature and carefully selected 399 published studies for consideration in formulating these evidence-based guidelines. The data were evaluated according to usefulness/efficacy of the intervention, level of evidence (eg, multiple randomized clinical trials, nonrandomized studies, case studies), and generalizability to women. Following is a summary of the important recommendations.

Lowering of LDL-C. The guidelines recommend initiating LDL-C-lowering therapy (preferably a statin) simultaneously with lifestyle therapy in high-risk women with an LDL-C level ≥100 mg/dL. In high-risk women with an LDL-C level <100 mg/dL, statin therapy is recommended unless contraindicated.

For women at intermediate risk, the guidelines advise LDL-C-lowering therapy if the LDL-C level is ≥130 mg/dL after a trial of lifestyle therapy. Low-risk women with 0 or 1 risk factor can be considered for LDL-C-lowering therapy if the LDL-C level is ≥190 mg/dL; however, in the presence of multiple risk factors (including the metabolic syndrome), low-risk women should be considered for statin therapy when the LDL-C level is ≥160 mg/dL.[3]

As with the ATP III guidelines, implementation of the AHA recommendations requires the use of clinical judgment, including consideration of lifestyle or emerging risk factors (eg, the metabolic syndrome) when determining the intensity of lipid-lowering therapy at any risk level. Once the patient's global risk has been assessed, it is important to calculate the amount of cholesterol lowering needed and to select a statin based on its average cholesterol-lowering effect at a given dose (see Chapter 5, Percentage Reduction in LDL).[25]

Modification of Triglyceride and HDL-C Levels

The AHA guidelines for women also address the need to modify high TG and low HDL-C levels. In high-risk women, if the HDL-C level is low or the non-HDL-C level is elevated, niacin (nicotinic acid; Niacor®, Niaspan® Extended Release) or a fibrate may be initiated concomitantly with statin therapy.

Although the AHA women's guidelines are based on a careful evaluation of clinical trial data, experts have questioned the initiation of niacin or a fibrate together with statin therapy. Both niacin and the fibrates have occasionally been associated with myopathy when administered as monotherapy, and the risk can increase in combination with a statin. The recently revised ATP III guidelines also consider the simultaneous initiation of an LDL-lowering drug and a fibrate or nicotinic acid to be a therapeutic option in high-risk patients with low HDL-C or high TG levels.

The prescribing information for each of the six available statins examines the risk for myopathy with these drug combinations. Precautions range from broad (eg, combined use of a statin plus niacin or fibrates should be avoided in general or unless the benefit will likely outweigh risk) to specific (ie, precise dosing instructions). In all cases, a statin at higher doses should never be combined with a fibrate. A formulation consisting of extended-release niacin and lovastatin is available (Advicor®), and the prescribing information for this product also warns about the possibility of myopathy and cautions physicians to weigh potential risks and benefits when contemplating its use.

Before niacin or a fibrate is prescribed with statin therapy, the patient's creatine kinase (CK) levels should be tested; thereafter, CK levels should be monitored periodically. Although interaction between a statin and niacin or a fibrate (particularly gemfibrozil [Lopid®]) tends to occur under circumstances in which such treatment should not be administered (eg, patients with renal insufficiency; those taking concomitant medications that interfere with clearance of statins; those older than 70 years), all patients must be apprised of the signs and symptoms of possible myopathy and told to contact their physician immediately if any occur (see Chapter 5, Drug Therapy/Fibric Acid Derivatives, for a summary of precautions).[26]

For women in the intermediate- and low-risk categories, the AHA women's guidelines recommend the use of niacin or fibrate therapy (to raise HDL-C or lower non-HDL-C) after the LDL-C goal has been reached.[3]

Adult Treatment Panel III Guidelines

The need for risk factor modification to reduce the incidence of CHD, the leading cause of death in women, is clear. However, the revised ATP III guidelines do not contain specific recommendations regarding women. The results of the Heart Protection Study (HPS) and the Prospective Study of Pravastatin in the Elderly at Risk (PROSPER), both of which enrolled substantial numbers of female subjects, suggest a need for further investigation to identify the most effective risk-reduction strategies in women. According to a prespecified subgroup analysis of women in the HPS (N=5,082; 25% of the study population), there was a statistically significant decrease of 19% in risk for a first vascular event with statin therapy vs placebo (compared with a 22% reduction in men). Although the number needed to treat (30) is reasonable, it is higher than that for men (17). In addition, the range of possible effect sizes implies some uncertainty about clinical significance. This may reflect a lack of statistical power to detect benefit in women. In PROSPER, the 3,000 women (52% of the study population) aged 70 to 82 years experienced a nonsignificant decrease of just 4% in the primary end point of CHD death, nonfatal MI, and fatal/nonfatal stroke with a statin vs placebo.

While adherence to the revised ATP III guidelines, whose recommendations are largely based on HPS and PROSPER, is critical for the reduction of cardiovascular risk overall, the above evidence suggests that specialized or individualized approaches may be needed in high-risk or elderly women.

Hormone Replacement Therapy

Hormone replacement therapy (HRT) is approved by the Food and Drug Administration (FDA) for the prevention of osteoporosis and for treatment of menopausal symptoms. It is not approved for lipid modification or CHD risk reduction. However, it once seemed to have promise as a cardiovascular intervention, largely because of epidemiologic data indicating that women using long-term estrogen replacement therapy (ERT) experienced a reduction in coronary events.[27,28] Studies also found that ERT decreased LDL-C and raised HDL-C levels depending on the type of estrogen, the regimen (ie, unopposed or in combination with a progesterone), the dose and dosing schedule, and the route of administration (oral, percutaneous, or transdermal).[29-40] However, recent epidemiologic and prospective clinical trials have cast substantial doubt on the cardioprotective role of HRT.

Randomized controlled trials. The Estrogen Replacement and Atherosclerosis (ERA) trial was a placebo-controlled study of estrogen alone or in combination with progesterone in 309 women with angiographically verified coronary disease.[41] This study found no change in the progression of atherosclerosis with either regimen, despite significant decreases in LDL-C and significant increases in HDL-C levels.[41] The placebo-controlled Postmenopausal Estrogen/Progestin Interventions (PEPI) trial evaluated the effect of estrogen alone or together with one of three progestational regimens on selected risk factors for coronary disease, including HDL-C and nonlipid risk factors.[42] Although unopposed estrogen was the most effective in raising HDL-C levels, it was associated with an increased risk for endometrial hyperplasia that would restrict its use to women without a uterus.

In the Heart and Estrogen/Progestin Replacement Study (HERS), 0.625 mg of conjugated equine estrogen (CEE) plus 2.5 mg medroxyprogesterone acetate produced no overall difference in cardiovascular outcomes compared with placebo, despite an 11% net reduction in LDL-C and a 10% net increase in HDL-C.[43] Moreover, in the first year of the study, women taking HRT experienced an increase in CHD events, although there were fewer events with HRT vs placebo in years 4 and 5. Consequently, the investigators concluded that HRT should not be initiated for the prevention of CHD, although it might be appropriate to continue therapy in women already receiving HRT.

Nurses' Health Study. The Nurses' Health Study, a prospective observational study begun in 1976 to investigate a time trend in risk for recurrent CHD, also found a trend of decreasing risk with increasing duration of HRT use (P=0.002).[44] Compared with women who had never used HRT, those receiving HRT for <1 year experienced an apparent 25% increase in coronary events, whereas ≥2 years of hormone use was associated with a 62% reduction in relative risk. Of the participants taking HRT, 53% used oral conjugated estrogen alone, 19% used oral conjugated estrogen plus oral progestin, and 28% used other forms of hormone therapy (primarily oral estradiol or transdermal estrogen).

Despite limited statistical power to detect differences in the effect of each regimen, there was no strong evidence of varying effects for estrogen alone or estrogen plus progestin with either short-term or longer-term use.[44]

Women's Health Initiative. Begun in 1992, the Women's Health Initiative (WHI) is a large clinical investigation program designed to evaluate the risks and benefits of strategies for preventing and controlling cardiovascular disease, cancer, and osteoporotic fractures in postmenopausal women aged 50 to 79 years. As part of the WHI, two parallel randomized, controlled trials of HRT were undertaken: CEE 0.625 mg/day vs placebo in women who had had a hysterectomy[45] and CEE 0.625 mg/day plus progestin (medroxyprogesterone acetate) 2.5 mg/day vs placebo in women with an intact uterus.[46]

Both trials were terminated early because of unfavorable outcomes in women receiving HRT. The CEE plus progestin trial (N=16,608) ended after 5.2 years (approximately 3 years early) because of a 26% increase in invasive breast cancer that just missed nominal statistical significance. Likewise, although the absolute number of CHD events with HRT vs placebo was small (164 vs 122), the relative increase was 29% (nominal 95% confidence interval [CI] 1.02 to 1.63). This consisted largely of a significant 32% increase in nonfatal MI.[46] Rates of stroke (primarily nonfatal) and venous thromboembolism were also significantly higher with HRT (41% and 100%, respectively). Overall, patients in the HRT group experienced a 22% increase in cardiovascular disease events (nominal 95% CI 1.09 to 1.36). In contrast, there was a 24% decrease in total osteoporotic fractures (nominal 95% CI 0.69 to 0.85) with HRT.[46]

The interventional phase of the CEE alone trial (N=10,739) was terminated in February 2004 (8 to 13 months early). Although HRT vs placebo produced no significant effect on the primary end point of CHD events, there was a significant 39% increase in stroke, consisting largely of nonfatal events, and a nonsignificant 33% increase in venous thromboembolism, an end point comprising deep vein thrombosis (DVT) and pulmonary embolism (only the DVT rate rose significantly). Overall, the total cardiovascular disease event rate increased 12% in women taking HRT (nominal 95% CI 1.01 to 1.24). For the primary safety outcome of invasive breast cancer, the event rate decreased by a nonsignificant 23% with HRT vs placebo, and total osteoporotic fractures decreased by 30% (nominal 95% CI 0.63 to 0.79).[45]

In each trial, only a small subset of women had a history of CHD or stroke, and the results in these participants did not differ significantly from the results in women without a similar history.[45,46]

Currently, the FDA recommends that physicians prescribe HRT at the smallest effective dose for the shortest possible time only to relieve certain postmenopausal symptoms (Table 6-1).[45] HRT should not be initiated or continued for the primary prevention of CHD, and its use for the prevention of osteoporotic fracture must be weighed against the substantially increased risk for cardiovascular disease.[46]

The WHI data have led to analyses that attempt to understand why the coronary outcomes differ from those of some observational studies, including the Nurses' Health Study. One possible reason is that, in the WHI trials, a majority of the participants were women older than 60 years. It has been suggested that the age at which HRT is initiated (<10 years vs ≥10 years from the start of menopause) may influence its effect on CHD risk.[47] This issue requires further study and discussion.

WISDOM. Following the termination of the WHI study, another major trial of HRT for the prevention of cardiovascular disease was also discontinued. The Women's International Study of Long Duration Oestrogen after Menopause (WISDOM) planned to involve 22,000 women aged 50 to 69 years from the United Kingdom, Australia, and New Zealand. Reasons for the discontinuation included difficulty in recruiting women into the study; strong evidence that HRT increases the risk for breast cancer; data indicating that HRT may increase cardiovascular risk in the short term, together with a lack of evidence that it is cardioprotective in the long term; and the expectation that results of the trial, which would not be available for another decade, would be unlikely to influence clinical practice.[48-50]

Elderly

Risk for CHD increases with age. Although clinical trial data concerning lipid-lowering therapy in the elderly have been scarce, evidence from the HPS indicates that treatment with a statin can reduce the risk for a first major vascular event by 23% in high-risk patients aged 65 to 69 years of age and by 18% in those aged 70 years or older. While the results in both age groups were statistically significant, the outcome in the over-70 population may be of uncertain clinical significance based on the range of possible effect sizes. This suggests that caution is needed regarding over-reliance on general cases. Accordingly, in the high-risk elderly, it is particularly important to individualize treatment decisions through the use of clinical judgment (see Elderly/Lipid Risk, Drug Therapy/Precautions, and Adult Treatment Panel III Guidelines, this chapter).

Although the results of PROSPER, examined below, further support pharmacologic intervention in older patients, particularly those with preexisting cardiovascular disease, the use of lipid-regulating medication in the elderly raises specific issues (Table 6-2). For example, older patients may require lower doses to achieve maximum reductions in LDL-C.

The modifiable risk factors for CHD in all populations include hypertension, dyslipidemia, habitual smoking, and physical inactivity. Although elderly people smoke less than their younger counterparts, they experience significant increases in hypertension and physical inactivity. Both groups have similar rates of hypercholesterolemia.[51]

Table 6-2: Special Considerations in Treating Dyslipidemia in the Elderly

Therapy	Consideration
Therapeutic lifestyle changes	Carefully individualize diet to ensure adequacy of nutrition. Initiate regular physical activity if appropriate.
Drug therapy	Use caution because elderly may be susceptible to adverse effects. Fibrates and statins are generally well tolerated. For statins, maximum reductions may be achieved with lower doses.
Resins	Associated constipation may be a particular problem. Absorption of other drugs may be decreased.
Statins	Most cases of severe myopathy have occurred in older patients, particularly those with coexisting disease (eg, renal insufficiency).

ULN = upper limit of normal

Lipid Risk

The utility of lipid screening in elderly patients has been questioned because of a reported weakened association between cholesterol and mortality in the oldest patient groups.[52] However, absolute rates of CVD increase with age, and the risk for CVD is a function of LDL-C level and

Therapy	Consideration
Drug therapy *(continued)*	
Fibrates	Incidence of gallstones may be increased; cholecystectomy carries more risk in older patients. Other possible side effects are gastrointestinal distress, impotence, and, in patients with renal insufficiency, myopathy.
Cholesterol absorption inhibitor	Limited data (pooled analysis) indicate a similar incidence of adverse events, including changes in liver function and creatine kinase elevations ≥10 x ULN, in combination with a statin vs statin therapy alone. Anecdotal evidence suggests that the safety and efficacy profile is similar in older and younger patients.
Nicotinic acid (niacin)	Common side effects may be more pronounced (eg, flushing, dry skin, dry mouth). Impaired glucose tolerance may be aggravated. Niacin raises uric acid levels and may precipitate gout.

duration of exposure.[53] Therefore, even moderately elevated LDL-C levels can be progressively detrimental over the course of a lifetime. Both the Los Angeles Veterans Hospital Administration Study, which placed men on a cholesterol-lowering diet, and the Stockholm Ischaemic Heart Disease Secondary Prevention Study, which used a combination of clofibrate and niacin in men and women, have produced data on the elderly in their respective cohorts. In these trials, younger and older participants experienced similar reductions in coronary morbidity.[51] A meta-analysis of the major statin trials has reported similar benefits in older and younger patients.[54] Furthermore, patients ≥65 years of age had the lowest number needed to treat to prevent one cardiovascular event.

These findings speak to an important concern regarding the elderly: while therapies that extend longevity by improving survival are important, is there a place for interventions that improve the quality of the remaining years of life by reducing the risk for developing potentially disabling diseases? Lipid modification may be useful for the latter objective through its effects on cardiovascular disease (see Risk Factors for Postmenopausal Women, this chapter).

Just as TG may be a prominent risk factor in women, its concentrations figure as strong indicators of coronary risk in elderly patients. The exact role of TG is unclear because of the dearth of clinical trials, but dietary change, weight loss, and exercise are essential in treating hypertriglyceridemia.[55]

Therapeutic Lifestyle Changes

The ATP III guidelines for treating high blood cholesterol recommend a comprehensive approach to reducing the risk for cardiovascular disease. This includes changing dietary habits that are atherogenic and engaging in more physical activity. Together, these modifications are called *therapeutic lifestyle changes* (Chapter 4). Dietary therapy should be approached carefully in elderly patients. Although data indicate that some elderly patients can manage lipid concentrations on diet alone,[56] dietary restrictions may represent a greater hardship or be inappropriate in the elderly population. Whereas some older patients may rigorously adhere to dietary restrictions, others may be unwilling to change lifelong habits. In either case, it is crucial that dietary recommendations in elderly patients satisfy their nutritional needs;

here, as elsewhere, counseling by a registered dietitian may provide an effective adjunct to regular medical care. The ATP III guidelines advocate referral to a registered dietitian or nutritionist for medical nutrition therapy, when appropriate, to enhance the effectiveness of dietary strategies or to improve patient compliance.

The Centers for Disease Control and Prevention and the American College of Sports Medicine have reported that 24% of elderly people are completely sedentary, while another 54% are inadequately active. Both men and women with CHD may benefit from exercise, although patients ≥75 years of age may not benefit as much as younger elderly patients. In a study by Hakim et al, activity as measured by the number of miles walked per day was correlated with mortality in 707 nonsmoking men, 61 to 81 years of age, from the Honolulu Heart Program. Patients who walked more than 2 miles a day had a lower all-cause mortality rate than those who walked less than 1 mile a day (23.8% vs 40.5%, P=0.001).[57] There was also a nonsignificant trend toward lower CHD rates among the more active patients. These data demonstrate an association between level of activity and mortality in patients physically capable of nonstrenuous exercise. When prescribing physical activity for the elderly, a principal concern is that it match the degree of functioning.

Drug Therapy

PROSPER. From the Scandinavian Simvastatin Survival Study (4S) through the HPS, major statin trials evaluated the effects of lipid lowering primarily in middle-aged subjects. To determine whether statin therapy is also safe and effective in older patients with, or at a high risk of developing, vascular disease, PROSPER randomized 5,804 men and women aged 70 to 82 years to treatment with pravastatin (Pravachol®) 40 mg/day or placebo.[58] Eligibility criteria included preexisting vascular disease or increased risk from smoking, hypertension, or diabetes; TC levels of 155 to 348 mg/dL; and TG levels <530 mg/dL. At entry, the mean

LDL-C level was 147 mg/dL. In this trial, the primary combined end point was CHD death, nonfatal MI, and fatal or nonfatal stroke. Secondary end points consisted of the coronary and cerebrovascular end points examined separately and the primary outcome evaluated in specified subgroups (ie, men, women, and subjects with and without preexisting vascular disease).[58]

In this 3-year study, LDL-C levels decreased approximately 33%.[3] There were significant reductions of 15% in the primary end point and 19% in the secondary end point of CHD death or nonfatal MI, but no reduction in the incidence of fatal or nonfatal stroke. A significant 24% reduction in CHD death was observed, but there was no significant reduction in nonfatal MI. The results of prespecified subgroup analyses indicate that subjects with established CVD experienced significant reductions in risk for a primary end-point event and for CHD death and nonfatal MI. Among patients without preexisting disease, there was a trend toward benefit in these outcomes and in transient ischemic attack, but the results were not statistically significant.

The lack of a favorable stroke outcome in PROSPER may be explained by inadequate statistical power. The authors also suggest that the trial's duration (3 years) may be insufficient for the emergence of a stroke benefit.[3] In the larger HPS (N=20,536; mean age of 64 years), however, there was a significant reduction of about one third in overall stroke risk by the end of the second year of therapy.[52] After 5 years of follow-up, HPS found a highly significant 25% reduction in stroke, with no evidence of a difference in the incidence of hemorrhagic stroke with statin therapy vs placebo. Among participants ≥65 years of age, the 5-year trend toward benefit was similar to the result in the total study population. Although age categories were not prespecified for the subgroup analysis of stroke alone, this favorable trend is reinforced by the results of prespecified subgroup analyses for the much larger numbers of first major vascular events, the primary end point.[59]

Adult Treatment Panel III Guidelines

Based on the results of HPS and PROSPER, together with data from other statin trials, the revised ATP III recommendations favor intensive LDL-lowering treatment in older adults with established cardiovascular disease. In elderly persons without established cardiovascular disease, the reliability of risk assessment is limited, in part because a lifetime of exposure to moderately elevated lipid levels may confer a higher degree of risk than might otherwise be inferred. Therefore, in addition to using the Framingham scoring system, physicians must exercise clinical judgment in deciding when to initiate intensive LDL-lowering therapy in older patients without a history of cardiovascular disease. In all elderly patients, issues of safety, tolerability, and patient preference must be carefully weighed.

Precautions

Elderly patients may be particularly susceptible to side effects associated with drug interventions. There is also a potential for increased toxicity in patients on multiple medications. The high cost of medication is often an additional issue in elderly patients with limited income.

In patients deemed appropriate for drug therapy, the statins and fibrates are generally well tolerated, although both carry a risk for side effects. For elderly patients, maximum reductions in lipid levels may be achieved with lower doses of statins. Although cholesterol absorption inhibitors have not been systematically studied in the elderly, anecdotal evidence suggests that a low-dose statin plus a cholesterol absorption inhibitor may be similarly effective in older and in younger patients. Bile-acid sequestrants (resins) may induce gastrointestinal complaints (eg, constipation) in the elderly and may prevent absorption of concomitant medications. Common side effects of nicotinic acid treatment may also be more pronounced in the elderly (Table 6-2).

It must be emphasized that treatment of the elderly should be individualized. While aggressive therapy may be appropriate for vigorous and otherwise healthy older patients, a conservative approach may be necessary for those who have other life-threatening diseases or multiple health problems. The clinician must assess how the elderly person's quality of life will be affected by diet or drug therapy.

Younger Patients

Multiple lines of evidence indicate that atherosclerotic disease begins to develop in childhood. The Pathobiological Determinants of Atherosclerosis in Youth (PDAY) study, an autopsy investigation conducted in young people who died of external causes (accidents, suicides, homicides), found correlations between established risk factors (eg, levels of LDL-C and TG, blood pressure, BMI, and cigarette smoking) and the presence and extent of atherosclerotic lesions 2 to 3 decades before the age when the clinical manifestations of atherosclerosis generally appear.[60] Other studies have found associations between CHD risk factors in children and adolescents and carotid intima-media thickness, a surrogate end point for cardiovascular disease, in the same individuals examined 2 to 3 decades later.[61,62] These data strongly suggest that preventive strategies initiated in the very young may hold great promise for reducing cardiovascular risk in adulthood.

Recently, the Geneva-based World Heart Federation warned that high-calorie diets and insufficient exercise (eg, hours spent in front of the computer and the television) are contributing to a surge in childhood obesity, and that children who are overweight or obese have a short-term risk of developing type 2 diabetes or cardiovascular disease (heart attack, stroke) before 65 years of age. According to a report by the London-based International Obesity Task Force, an estimated 10% of children (at least 155 million worldwide) are overweight or obese.[63]

Screening

In 2003, the AHA issued guidelines for the primary prevention of atherosclerotic cardiovascular disease beginning in childhood.[64] Citing a growing body of evidence to support the safety and effectiveness of early intervention, the authors of these guidelines recommend that screening for cardiovascular disease start at a very young age.[64]

Pediatric care providers should regularly update a child's family history (parents, grandparents, aunts, uncles) to identify a history of obesity, hypertension, dyslipidemia, diabetes, cigarette smoking, and premature cardiovascular disease. At every visit, the child's height, weight, BMI, and (after age 3) blood pressure should be determined. Diet and physical activity should also be assessed, as well as cigarette smoking in children ≥9 years of age.

If the family history is positive for cardiovascular risk factors, or if the family history is unknown, fasting plasma lipid levels should be screened regularly in children older than 2 years. Total cholesterol and LDL-C levels of >170 mg/dL and >110 mg/dL, respectively, are considered borderline (>200 mg/dL and >130, respectively, are elevated). Triglyceride levels >150 mg/dL and HDL-C levels <35 mg/dL also warrant attention, as does a BMI >85th percentile.

The guidelines do not specifically address the issue of familial hypercholesterolemia (FH), an underdetected genetic disorder that is associated with an elevated risk for premature CHD. Heterozygous FH affects approximately 1 in 500 people; the more severe homozygous form occurs in about 1 in 1 million people.[65] Experts have not yet reached a consensus on the most effective method for identifying FH: DNA testing or family screening.[66-68] Until consensus guidelines are available, physicians should carefully update the family histories of their pediatric patients and assess fasting plasma lipid levels on a regular basis, as recommended by the AHA guidelines for primary cardiovascular disease prevention in children.

Treatment

In pediatric patients with an LDL-C level ≥130 mg/dL (≥100 mg/dL in children with diabetes), lifestyle recommendations should be implemented in consultation with a trained dietitian. These recommendations include <7% of calories from saturated fat, <200 mg cholesterol per day, the addition of dietary fiber in an age-appropriate manner, and an emphasis on weight reduction and increased physical activity.[64] If the LDL-C remains elevated, secondary causes should also be investigated (eg, thyroid-stimulating hormone tests, liver function tests, renal function tests, urinalysis).

For children or adolescents without other cardiovascular risk factors, the AHA guidelines recommend consideration of pharmacotherapy if LDL-C levels remain >190 mg/dL. When other risk factors or a strong family history of premature cardiovascular disease is present, pharmacotherapy should be considered if LDL-C levels remain >160 mg/dL.

In 1998, the American Academy of Pediatrics (AAP) issued guidelines stating that drug therapy should be considered only for children ≥10 years of age if, after 6 to 12 months of dietary modification, the LDL-C level remains ≥190 mg/dL (≥160 mg/dL in those with a family history of premature cardiovascular disease or ≥2 uncontrollable risk factors).[69] When drug therapy is deemed appropriate in children or adolescents, the AAP recommends the use of bile acid sequestrants and cautions against the routine use of other agents.

Since the AAP guidelines were published, two statins—pravastatin and atorvastatin (Lipitor®)—have been approved for the treatment of heterozygous FH in children ≥8 years old and ≥10 years old, respectively, who meet specific criteria.[70,71]

In all cases, drug therapy should be prescribed in consultation with a physician who specializes in the treatment of lipid disorders in children.

Diabetes Patients

Type 1 and type 2 diabetes are CHD risk equivalents. Coronary heart disease occurs more frequently and at a younger age among diabetic individuals than in the general population and is the cause of death for more than half of the adult diabetic population.

Diabetic patients, particularly those with type 2 diabetes, are at high risk because of elevated levels of TG and VLDL particles, as well as low levels of HDL-C. Although patients with type 2 diabetes usually have LDL-C levels that are not significantly different from those in nondiabetic individuals, they tend to produce small, dense LDL, which is more vulnerable to oxidation.[72,73] The LDL particles of patients with type 2 diabetes also have a higher TG content than those of normoglycemic patients.[74] Other mechanisms that may promote heart disease in diabetes patients include the glycation of arterial wall proteins, increased LDL oxidation, microvascular disease of the vasa vasorum, change in cellular function, promotion of thrombogenesis, and the development of hypertension and renal disease.

Hypertriglyceridemia and Low HDL-C

Poorly controlled diabetes is characterized by hypertriglyceridemia and low HDL-C levels, the most common pattern of dyslipidemia in patients with type 2 diabetes.[75] The condition is a result of both increased hepatic VLDL-TG production and defective VLDL clearance. It is generally thought that insulin deficiency leads to the impairment of plasma TG clearance. Evidence also suggests that insulin fails to suppress VLDL in the liver. The increased entry of free fatty acid into the liver may be another contributing factor.[76]

In a Finnish study, lipids and lipoprotein fractions were measured for 7 years in 313 patients with type 2 diabetes. Elevated TG levels increased the risk for CHD death and all CHD events twofold. Low HDL-C was the most powerful risk factor because it increased the risk for CHD death fourfold and the risk for all CHD events twofold.[73]

Independent risk factors (eg, obesity) may create a more atherogenic environment for diabetic individuals, as they do among people without diabetes.[76] Studies suggest that TC is a risk factor in both diabetic and nondiabetic patients. However, TC concentration may not reflect the distribution of cholesterol within the lipid fractions in diabetic patients; while LDL-C is the main contributor to TC in nondiabetic patients, excess VLDL-C may significantly contribute to TC concentration in diabetic dyslipidemia. Consequently, TC and VLDL-TG are stronger risk factors for CHD in patients with type 2 diabetes than in nondiabetic patients.[74]

Renal Disease and Screening

A strong link between renal disease and CHD exists for patients with type 1 diabetes. In a study of 658 patients with type 1 diabetes, 37 died in the first 4 years of follow-up. Twenty-two deaths (59%) were caused by CHD, while 16% were the result of diabetic coma. Eighty-one percent (81%) of the patients who died from CHD causes also had renal disease. Although only 41% of the deceased patients showed evidence of CHD during their last biennial examination, 54% had evidence of lower extremity arterial disease.

A multivariate analysis has suggested that smoking history, TG levels, and total platelet count are independent predictors of mortality, while LDL-C levels are the best indicators of CHD mortality. The study highlights the need for more intensive screening for cardiovascular and renal disease in patients with type 1 diabetes, followed by correction of cardiovascular risk factors. Lipid management may reduce the development of both diseases.[76]

Clinical Trials

Of the 20,536 HPS participants, 5,963 were known to have diabetes. In these patients, data from a prespecified subgroup analysis indicate that simvastatin (Zocor®) 40 mg decreased LDL-C levels by approximately 39 mg/dL from an average of 128 mg/dL at baseline. Reductions in LDL-C were similar to those in nondiabetic subjects, irrespective of presenting feature (eg, with/without prior CHD or cardiovascular disease, age <65 or ≥65 years, LDL-C <116 or ≥116 mg/dL). Corresponding to the reduction in LDL-C, the risk for major coronary events decreased by 27%

(P <0.0001) vs placebo; this included reductions of 20% in coronary mortality (P=0.02) and 37% in nonfatal MI (P=0.0002). Simvastatin also produced significant decreases of 24% and 17% in the risk for stroke and for revascularizations, respectively, in diabetic subjects. These results were not significantly different from those in subjects without diabetes, and the proportional reductions in risk appeared to be independent of the time since diagnosis of diabetes.[77] The investigators conclude that statin therapy should be considered routinely for all diabetic patients at sufficiently high risk for any major vascular event, regardless of initial cholesterol concentrations.[77]

The Collaborative Atorvastatin Diabetes Study (CARDS)[78,79] provides further evidence concerning the efficacy of statin therapy in patients with diabetes. Participants in this randomized, double-blind, placebo-controlled study (N=2,838; mean age of 62 years) had type 2 diabetes (a mean of 7.9 years from diagnosis at baseline); no clinical evidence of cardiovascular disease (excluding hypertension); LDL-C and TG levels ≤160 mg/dL and ≤600 mg/dL, respectively; and at least one other risk factor (ie, hypertension, retinopathy, micro-/macroalbuminuria, or current smoking). The primary end point was the time from randomization to occurrence of the first major cardiovascular event (CHD death, nonfatal MI, unstable angina, resuscitated cardiac arrest, revascularization, stroke). In this study, participants receiving atorvastatin 10 mg/day had a significant 37% reduction in the composite primary end point (P=0.001), which included a 48% reduction in stroke. The effect was consistent regardless of age, sex, and baseline lipid levels or risk factors. A 27% reduction in all-cause mortality just missed statistical significance. During the 4-year follow-up period, the median LDL-C was approximately 77 mg/dL with atorvastatin; throughout the study, ≥75% of subjects receiving atorvastatin had an LDL-C level <95 mg/dL, and ≥25% had a level <64 mg/dL. Because the significant difference in favor of atorvastatin met a prespecified criterion for early termination,

Table 6-3: Categories of Risk According to Baseline Lipid Levels (mg/dL) in Type 2 Diabetes

Risk	LDL-C	HDL-C	Triglyceride
Higher	≥130	<40	>400
Borderline	100-129		150-400
Lower	<100	>50	<150

HDL-C = high-density lipoprotein cholesterol
LDL-C = low-density lipoprotein cholesterol

American Diabetes Association, *Diabetes Care* 2004;27(suppl 1): S68-S71.

CARDS was stopped approximately 2 years before the scheduled completion date.[78,79]

By demonstrating that statin therapy can safely reduce cardiovascular risk in type 2 diabetes patients at the lower end of the cholesterol distribution range, these results support the importance of assessing total risk, with target LDL-C levels to be determined by absolute risk and the safety/tolerability of drug therapy. At the same time, the absence of heterogeneity in treatment effect across a variety of prespecified subgroups suggests that the treatment benefit could be extended to diabetes patients who do not have the additional risk factors of the participants in CARDS.

Further information on reducing cardiovascular risk in type 2 diabetes patients will be provided by the Action to Control Cardiovascular Risk in Diabetes (ACCORD) trial, a large clinical investigation sponsored by the National Heart, Lung, and Blood Institute. Scheduled for completion in 2010, ACCORD will evaluate three intensive strategies: lowering of blood glucose; blood pressure reduction in the context of good blood glucose control; and lipid modification with a statin and a fibrate.[80]

Table 6-4: ADA Treatment Recommendations Based on LDL-C (mg/dL) in Adults With Diabetes

Medical Nutrition Therapy		*Drug Therapy*	
Initiation level	**LDL-C goal**	**Initiation level**	**LDL-C goal**
≥100	<100	≥130 (>100 w/CVD)	<100

Caveats: (1) Medical nutrition therapy should be attempted before starting pharmacologic therapy; however, in patients with clinical CVD and LDL-C >100 mg/dL, pharmacologic therapy and lifestyle interventions should be initiated simultaneously. (2) In diabetes patients older than 40 years and with a TC level ≥135 mg/dL, statin therapy to decrease LDL-C roughly 30%, regardless of baseline levels, may be appropriate.

CVD = cardiovascular disease; LDL-C = low-density lipoprotein cholesterol; TC = total cholesterol

American Diabetes Association, *Diabetes Care* 2004;27(suppl 1): S68-S71.

Treatment

Table 6-3 describes the categories of risk by lipoprotein levels in patients with type 2 diabetes. Because of the frequent fluctuations in lipid concentrations observed in this group, the ADA recommends measuring LDL-C, HDL-C, TC, and TG every year in adult patients.[72] Children with diabetes may be considered for lipoprotein analysis after 2 years of age. The values given in the lower risk category are considered to be ideal for diabetic patients, with perhaps an emphasis on higher HDL-C in diabetic women. (Women in general tend to have higher HDL-C concentrations than men regardless of risk.)

Table 6-5: Drug Considerations in Diabetic Patients

Drug	Consideration
Resins	Resins are not recommended as first-line therapy in diabetic patients because of the potential to exacerbate hypertriglyceridemia.
Statins	Statins can be used to lower elevated LDL-C in diabetic patients; some statins have also demonstrated appreciable effects on TG concentration. The risk for myopathy is low, and, in general, statins do not worsen glucose control. Concomitant therapy with a statin and gemfibrozil is generally not recommended because of an increased risk for muscle toxicity.
Fibrates	Because of their TG-lowering and HDL-C raising effects, the fibrates are considered a good choice in diabetic patients, who often present with elevated TG and low HDL-C. Fibrates should be used cautiously or not at all in patients who have diabetic nephropathy with renal insufficiency because of the risk for myopathy. Fibrates should not be combined with statins in diabetic patients.

Experts who treat diabetes patients differ in their views of when to begin LDL-C-lowering therapy in relation to the initiation of lifestyle interventions. Table 6-4 presents the ADA guidelines,[72] which recommend pharmacologic therapy after lifestyle interventions alone have proved suboptimal. However, patients with preexisting cardiovascular disease and an LDL-C level >100 mg/dL should be

Drug	Consideration
Cholesterol absorption inhibitor	Based on limited data (post-hoc analysis), a statin plus ezetimibe appears to be well tolerated in patients with type 2 diabetes. Reductions in LDL-C, TG, and non-HDL-C were greater with statin plus ezetimibe vs statin alone. Patients with and without type 2 diabetes experienced similar benefit.
Nicotinic acid	Lipid-lowering doses of nicotinic acid are generally not recommended as first-line treatment in diabetic patients because nicotinic acid increases insulin resistance, hyperglycemia, and hyperinsulinemia. In diabetic patients with refractory dyslipidemia, nicotinic acid may be considered, but patients should be monitored very closely.

HDL-C = high-density-lipoprotein cholesterol,
LDL-C = low-density-lipoprotein cholesterol, TG = triglyceride

started on drug and behavioral therapy at the same time. The ADA also states that patients aged >40 years who have a TC level ≥135 mg/dL can be considered for statin therapy to lower LDL-C by about 30%, regardless of baseline levels.[72] This implies that such patients may also be candidates for the simultaneous initiation of pharmacologic and lifestyle interventions.

First-line therapy for hypertriglyceridemia consists of improved glycemic control and lifestyle intervention. Although the ADA guidelines allow for the possible use of fibrates or nicotinic acid to either raise HDL-C or lower TG, a fibrate should not be combined with a statin in diabetic patients, and nicotinic acid can worsen glycemic control.[72] If further cholesterol lowering is needed, a cholesterol absorption inhibitor added to statin therapy may be a desirable option (Table 6-5).

The European Atherosclerosis Society, the ATP III, the European Task Force, the ADA, and the International Lipid Information Bureau all advise aggressive risk factor intervention for the diabetic population.

Patients with type 1 diabetes often present with acceptable lipid profiles if there is good glycemic control.

Dietary Therapy

Weight reduction, a low-fat diet, regular physical activity, and glycemic control are fundamental to lipid management in patients with diabetes. Dietary therapy must be tailored to the individual, and consultation with a registered dietitian may help formulate an appropriate strategy. Because carbohydrates can raise TG levels, the source and the amount of daily intake are important, with an emphasis on soluble fiber as an important consideration in maintaining glycemic control.

Adult Treatment Panel III Guidelines

In diabetes patients with cardiovascular disease, the revised ATP III guidelines favor initiating statin therapy regardless of LDL-C level, with the optional goal of <70 mg/dL. For diabetes patients without established CHD, however, therapeutic choices may vary. Many have a cardiovascular risk level equivalent to that of nondiabetic patients with established CHD. Reasons can include older age and the presence of multiple risk factors. In this high-risk group, the HPS results support reducing LDL-C to <100 mg/dL. However, the HPS also found that diabetes patients without cardiovascular disease experienced only marginally significant benefit with statin therapy if their baseline LDL-C level was <116 mg/dL. Therefore, clinical judgment is needed in deciding whether to initiate LDL-lowering therapy in diabetes patients without cardiovascular disease who present with an LDL-C level <100 mg/dL.

Some diabetes patients may be at moderately high risk (10% to 20%) because of younger age or absence of other risk factors. This category of patient was not studied in the HPS. In diabetes patients considered to be at lower risk, LDL-lowering therapy might not be initiated if the baseline LDL-C level is <130 mg/dL. However, a portion of diabetes patients with a 10-year risk under 20% may nevertheless be considered high risk because they have a poor prognosis once CHD becomes evident.

Maximal therapeutic lifestyle changes are needed in all diabetes patients, but clinical judgment in selected cases can help determine if there is a need for LDL-lowering therapy.

Acute Coronary Syndromes

Adult Treatment Panel III Guidelines

Chapter 1 examined the results of clinical trials in patients with acute coronary syndromes (ACS). Based on evidence from the Myocardial Ischemia Reduction with Aggressive Cholesterol Lowering (MIRACL) trial and the Pravastatin or Atorvastatin Evaluation and Infection Therapy (PROVE-IT) trial, ATP III advises consideration of intensive therapy for all patients hospitalized with an ACS. Moreover, evidence from PROVE-IT supports efforts to achieve the optional LDL-C goal of <70 mg/dL.[81] In selecting the LDL-lowering drug and the dosage, physicians should be guided in part by the patient's LDL-C level within 24 hours of admission. Acute coronary syndrome patients with relatively low LDL-C levels may be able to reach the optional goal of <70 mg/dL with standard-dose therapy; for those with higher baseline levels, intensive statin therapy or the combination of a lower-dose statin with ezetimibe (Zetia®), a resin, or nicotinic acid may be appropriate (see Chapter 5, Drug Therapy, for information on combination regimens). During follow-up, treatment regimens can be modified.

Safety and tolerability are essential considerations when assessing therapeutic options. In PROVE-IT, for example, there were significantly more hepatic adverse effects with high-dose atorvastatin than with standard-dose pravastatin. The authors caution that patients in clinical practice may be less able to tolerate high-dose statin therapy because they may have more coexisting conditions than the study participants.[81]

The ATP III guideline update was issued before publication of the Aggrastat-to-Zocor (A-to-Z) trial results. Although A-to-Z showed a trend toward clinical benefit, intensive statin therapy failed to produce significant event reduction. This outcome may reflect the importance of drug and dosage selection, as well as timing, in the treatment of post-ACS patients (see Chapter 1).

Despite the possibility that high-dose statin therapy may exert an anti-inflammatory effect during the period immediately following an ACS, experts generally prefer to begin with standard doses, monitor for safety/efficacy, and cautiously increase the dose, as appropriate. In real-life settings, comorbidities and other complications necessitate extreme caution and a commitment to careful monitoring when deciding whether to prescribe high-dose therapy.

The hospitalization period provides an opportunity to educate patients about the importance of lipid-lowering therapy. After an individual leaves the hospital, there may be gaps in follow-up, as well as a decrease in motivation to comply with treatment. Therefore, even if the prevention of an early recurrent event is uncertain, the clear long-term benefits of lipid lowering in this high-risk population may warrant initiation of therapy before discharge. This is consistent with a joint recommendation issued by the AHA and the American College of Cardiology.[81]

References

1. Expert Panel on Diabetes, Evaluation, and Treatment of High Blood Cholesterol in Adults: Executive Summary of the Third Report of the National Cholesterol Education Program (NCEP) Expert Panel on Detection, Evaluation, and Treatment of High Blood Cholesterol in Adults (Adult Treatment Panel III). *JAMA* 2001;285:2486-2497.

2. Grundy SM, Cleeman JI, Merz CN, et al: Implications of recent clinical trials for the National Cholesterol Education Program Adult Treatment Panel III guidelines. *Circulation* 2004;110:227-239.

3. Mosca L, Appel LJ, Benjamin EJ, et al: Evidence-based guidelines for cardiovascular disease prevention in women. *Circulation* 2004;109:672-693.

4. American Heart Association: Heart disease and stroke statistics–2004 update. Available at: http://www.americanheart.org/presenter.jhtml?identifier=3000090. Accessed September 30, 2004.

5. Manhas A, Farmer JA: Hypolipidemic therapy and cholesterol absorption. *Curr Atheroscler Rep* 2004;6:89-93.

6. Stangl V, Baumann G, Stangl K: Coronary atherogenic risk factors in women. *Eur Heart J* 2002;23:1738-1752.

7. Dunn NR, Faragher B, Thorogood M, et al: Risk of myocardial infarction in young female smokers. *Heart* 1999;82:581-583.

8. Rosenberg L, Palmer JR, Rao S, et al: Low-dose oral contraceptive use and the risk of myocardial infarction. *Arch Intern Med* 2001;161:1065-1070.

9. Croft P, Hannaford PC: Risk factors for acute myocardial infarction in women: evidence from the Royal College of General Practitioners' oral contraception study. *Br Med J* 1989;298:165-168.

10. Karrar A, Sequeira W, Block JA: Coronary artery disease in systemic lupus erythematosus: a review of the literature. *Semin Arthritis Rheum* 2001;30:436-443.

11. Colhoun HM, Rubens MB, Underwood SR, et al: The effect of type 1 diabetes mellitus on the gender difference in coronary artery calcification. *J Am Coll Cardiol* 2000;36:2160-2167.

12. Lloyd CE, Kuller LH, Ellis D, et al: Coronary artery disease in IDDM. Gender differences in risk factors but not risk. *Arterioscler Thromb Vasc Biol* 1996;16:720-726.

13. Pyorala K: Diabetes and coronary artery disease: what a coincidence? *J Cardiovasc Pharmacol* 1990;16(suppl 9):S8-S14.

14. Beard CM, Fuster V, Annegers JF: Reproductive history in women with coronary heart disease. A case-control study. *Am J Epidemiol* 1984;120:108-114.

15. Wild S, Pierpoint T, McKeigue P, et al: Cardiovascular disease in women with polycystic ovary syndrome at long-term follow-up: a retrospective cohort study. *Clin Endocrinol (Oxf)* 2000;52:595-600.

16. Talbott E, Guzick D, Clerici A, et al: Coronary heart disease risk factors in women with polycystic ovary syndrome. *Arterioscler Thromb Vasc Biol* 1995;15:821-826.

17. Redberg RF: Coronary artery disease in women: understanding the diagnostic and management pitfalls. Medscape General Medicine 199;1(3). Posted 10/29/1998. Available at: http://www.medscape.com/viewarticle/408890. Accessed October 1, 2004.

18. Sutton-Tyrrell K, Lassila HC, Meilahn E, et al: Carotid atherosclerosis in premenopausal and postmenopausal women and its association with risk factors measured after menopause. *Stroke* 1998;29:1116-1121.

19. Stefanick ML, Mackey S, Sheehan M, et al: Effects of diet and exercise in men and postmenopausal women with low levels of HDL cholesterol and high levels of LDL cholesterol. *N Engl J Med* 1998;339:12-20.

20. Reeder BA, Senthilselvan A, Despres JP, et al: The association of cardiovascular disease risk factors with abdominal obesity in Canada. Canadian Heart Health Surveys Research Group. *CMAJ* 1997;157(suppl 1):S39-S45.

21. Heim DL, Holcomb CA, Loughin TM: Exercise mitigates the association of abdominal obesity with high-density lipoprotein cholesterol in premenopausal women: results from the third National Health and Nutrition Examination Survey. *J Am Diet Assoc* 2000; 100:1347-1353.

22. DiPietro L, Katz LD, Nadel ER: Excess abdominal adiposity remains correlated with altered lipid concentrations in healthy older women. *Int J Obes Relat Metab Disord* 1999;23:432-436.

23. Rich-Edwards JW, Manson UE, Hennekens CH, et al: The primary prevention of coronary heart disease in women. *N Engl J Med* 1995;332:1758-1766.

24. Barrett-Connor E, Giardina EG, Gitt AK, et al: Women and heart disease: the role of diabetes and hyperglycemia. *Arch Intern Med* 2004;164:934-942.

25. Roberts WC: The rule of 5 and the rule of 7 in lipid-lowering by statin drugs. *Am J Cardiol* 1997;80:106-107.

26. Jones PH: Statin plus fibrate combination therapy. Role of fibrates in lipid treatment. Lipids Online [Web site]. Available at: http://www.lipidsonline.org/slides/slide01.cfm?q=fibrates&dpg=54. Posted February 1, 2000. Accessed June 30, 2004.

27. Stampfer MJ, Colditz GA: Estrogen replacement therapy and coronary heart disease: a quantitative assessment of the epidemiologic evidence. *Prev Med* 1991;20:47-63.

28. Ettinger B, Friedman GD, Bush T, et al: Reduced mortality associated with long-term postmenopausal estrogen therapy. *Obstet Gynecol* 1996;87:6-12.

29. Davidson MH, Maki KC, Marx P, et al: Effects of continuous estrogen and estrogen-progestin replacement regimens on cardiovascular risk markers in postmenopausal women. *Arch Intern Med* 2000;106:3315-3325.

30. Itoi H, Minakami H, Iwasaki R, et al: Comparison of the long-term effects of oral estriol with the effects of conjugated estrogen on serum lipid profile in early menopausal women. *Maturitas* 2000;36:217-222.

31. Sanada M, Nakagawa H, Kodama I, et al: Three-year study of estrogen alone versus combined with progestin in postmenopausal women with or without hypercholesterolemia. *Metabolism* 2000;49:784-789.

32. Owens D, Collins PB, Johnson A, et al: Lipoproteins and low-dose estradiol replacement therapy in post-menopausal type 2 diabetic patients: the effect of addition of norethisterone acetate. *Diabet Med* 2000;17:308-315.

33. Tonstad S, Os I: Does postmenopausal hormone replacement therapy have a place in treatment of hyperlipidemia? [in Norwegian] *Tidsskr Nor Laegeforen* 2000;120:923-926.

34. Weintraub M, Grosskopf I, Charach G, et al: Hormone replacement therapy enhances postprandial lipid metabolism in postmenopausal women. *Metabolism* 1999;48:1193-1196.

35. Wakatsuki A, Ikenoue N, Sagara Y: Effect of estrogen on the size of low-density lipoprotein particles in postmenopausal women. *Obstet Gynecol* 1997;90:22-25.

36. Kim CJ, Min YK, Ryu WS, et al: Effect of hormone replacement therapy on lipoprotein(a) and lipid levels in postmenopausal

women. Influence of various progestogens and duration of therapy. *Arch Intern Med* 1996;156:1698-1700.

37. Soma MR, Meschia M, Bruschi F, et al: Hormonal agents used in lowering lipoprotein(a). *Chem Phys Lipids* 1994;67-68:345-350.

38. Basdevant A, DeLignieres B, Guy-Grand B: Differential lipemic and hormonal responses to oral and parenteral 17 beta-estradiol in postmenopausal women. *Am J Obstet Gynecol* 1983;147:77-81.

39. Rodriguez-Aleman F, Torres JM, Cuadros JL, et al: Effect of estrogen-progestin replacement therapy on plasma lipids and lipoproteins in postmenopausal women. *Endocr Res* 2000;26:263-273.

40. Bongard V, Ferrieres J, Ruidavets JB, et al: Transdermal estrogen replacement therapy and plasma lipids in 693 French women. *Maturitas* 1998;30:265-272.

41. Herrington DM, Reboussin DM, Brosnihan KB, et al: Effects of estrogen replacement on the progression of coronary-artery atherosclerosis. *N Engl J Med* 2000;343:522-529.

42. Effects of estrogen or estrogen/progestin regimens on heart disease risk factors in postmenopausal women. The postmenopausal Estrogen/Progestin Interventions (PEPI) trial. The Writing Group for the PEPI Trial. *JAMA* 1995;273:199-208.

43. Hulley S, Grady D, Bush T, et al: Randomized trial of estrogen plus progestin for secondary prevention of coronary heart disease in postmenopausal women. Heart and Estrogen/Progestin Replacement Study (HERS) Research Group. *JAMA* 1998;280: 605-613.

44. Grodstein F, Manson JE, Stampfer MJ: Postmenopausal hormone use and secondary prevention of coronary events in the Nurses' Health Study. *Ann Intern Med* 2001;135:1-8.

45. Anderson GL, Limacher M, Assaf AR, et al: Effects of conjugated equine estrogen in postmenopausal women with hysterectomy: the Women's Health Initiative randomized controlled trial. *JAMA* 2004;291:1701-1712.

46. Risks and benefits of estrogen plus progestin in healthy postmenopausal women. Principal results from the Women's Health Initiative randomized controlled trial. *JAMA* 2002;288:321-333.

47. Grodstein F, Manson JE, Stampfer MJ, et al: The discrepancy between observational studies and randomized trials of menopausal hormone therapy [letter]. *Ann Intern Med* 2004;140:764-765.

48. UK researchers halt research into hormone replacement therapy. October 23, 2002. Available at: http://www.arabia.com/afp/tech/health/article/print/english/0,11868,316981,00.html. Accessed October 29, 2002.

49. HRT trial stopped early. October 24, 2002. Available at: http://uk.news.yahoo.com/021024/103/44xz.html. Accessed November 4, 2002.

50. Huge HRT trial is abandoned: new evidence clouds future of menopause drug. October 24, 2002. Available at: http://www.heart1.com/news/newsfeed.cfm/1254. Accessed November 4, 2002.

51. Williams MA: Cardiovascular risk-factor reduction in elderly patients with cardiac disease. *Phys Ther* 1996;76:469-480.

52. Schatz IJ, Masaki K, Yano K, et al: Cholesterol and all-cause mortality in elderly people from the Honolulu Heart Program: a cohort study. *Lancet* 2001;358:351-355.

53. Gotto AM Jr, Kuller LH: Eligibility for lipid-lowering drug therapy in primary prevention: how do the Adult Treatment Panel II and Adult Treatment Panel III guidelines compare? *Circulation* 2002;105:136-139.

54. La Rosa JC, He J, Vupputuri S: Effect of statins on risk of coronary disease: a meta-analysis of randomized controlled trials. *JAMA* 1999;282:2340-2346.

55. Malloy MJ, Kane JP: A risk factor for atherosclerosis: triglyceride-rich lipoproteins. *Adv Intern Med* 2001;47:111-136.

56. Fleming RM, Ketchum K, Fleming DM, et al: Treating hyperlipidemia in the elderly. *Angiology* 1995;46:1075-1083.

57. Hakim AA, Petrovitch H, Burchfiel CM, et al: Effects of walking on mortality among nonsmoking retired men. *N Engl J Med* 1998;338:94-99.

58. Shepherd J, Blauw GJ, Murphy MB, et al: Pravastatin in elderly individuals at risk of vascular disease (PROSPER): a randomised controlled trial. *Lancet* 2002;360:1623-1630.

59. Collins R, Armitage J, Parish S, et al: Effects of cholesterol-lowering with simvastatin on stroke and other major vascular events in 20,536 people with cerebrovascular disease or other high-risk conditions. *Lancet* 2004;363:757-767.

60. McGill HC Jr, McMahan CA, Zieske AW, et al: Associations of coronary heart disease risk factors with the intermediate lesion of atherosclerosis in youth. The Pathobiological Determinants of

Atherosclerosis in Youth (PDAY) Research Group. *Arterioscler Thromb Vasc Biol* 2000;20:1998-2004.

61. Li S, Chen W, Srinivasan SR, et al: Childhood cardiovascular risk factors and carotid vascular changes in adulthood: the Bogalusa Heart Study. *JAMA* 2003;290:2271-2276.

62. Raitakari OT, Juonala M, Kahonen M, et al: Cardiovascular risk factors in childhood and carotid artery intima-media thickness in adulthood: the Cardiovascular Risk in Young Finns Study. *JAMA* 2003;290:2277-2283.

63. Risk of future heart attack higher in hefty kids. Monday, September 20, 2004. CNN.com: International Edition. Available at: http://www.cnn.com/2004/HEALTH/conditions/09/20/heart.kids.reut/. Accessed October 4, 2004.

64. Kavey RE, Daniels SR, Lauer RM, et al: American Heart Association guidelines for primary prevention of atherosclerotic cardiovascular disease beginning in childhood. *Circulation* 2003;107: 1562-1566.

65. Rifkind BM, Schucker B, Gordon DJ: When should patients with heterozygous familial hypercholesterolemia be treated? [editorial]. *JAMA* 1999;281:180-181.

66. Umans-Eckenhausen MA, Defesche JC, Sijbrands EJ, et al: Review of first 5 years of screening for familial hypercholesterolaemia in the Netherlands. *Lancet* 2001;357:165-168.

67. Schuster H: Risk assessment and strategies to achieve lipid goals: lessons from real-world clinical practice. *Am J Med* 2004;116(suppl 6A):26S-30S.

68. Marks D, Wonderling D, Thorogood M, et al: Cost effectiveness analysis of different approaches of screening for familial hypercholesterolaemia. *BMJ* 2002;324:1303-1308.

69. American Academy of Pediatrics Committee on Nutrition: Cholesterol in childhood. *Pediatrics* 1998;101:141-147.

70. Pravachol® (pravastatin sodium) tablets [package insert]. Princeton, NJ, Bristol-Myers Squibb Company, 2003.

71. Lipitor® (atorvastatin calcium) tablets [package insert]. New York, Parke-Davis Division of Pfizer Inc, 2003.

72. Haffner SM: Dyslipidemia management in adults with diabetes. *Diabetes Care* 2004;27(suppl 1):S68-S71.

73. Nesto R: CHD: a major burden in type 2 diabetes. *Acta Diabetol* 2001;38(suppl 1):S3-S8.

74. Laakso M: Lipids and lipoproteins as risk factors for coronary heart disease in non-insulin-dependent diabetes mellitus. *Ann Med* 1996;28:341-345.

75. Yoshino G, Hirano T, Kazumi T: Dyslipidemia in diabetes mellitus. *Diabetes Res Clin Pract* 1996;33:1-14.

76. Portuese EI, Kuller L, Becker D, et al: High mortality from unidentified CVD in IDDM: time to start screening? *Diabetes Res Clin Pract* 1995;30:223-231.

77. MRC/BHF Heart Protection Study of cholesterol-lowering with simvastatin in 5,963 people with diabetes: a randomised placebo-controlled trial. *Lancet* 2003;361:2005-2016.

78. Colhoun HM, Betteridge DJ, Durrington PN, et al: Primary prevention of cardiovascular disease with atorvastatin in type 2 diabetes in the Collaborative Atorvastatin Diabetes Study (CARDS): multicentre randomised placebo-controlled trial. *Lancet* 2004; 364:685-696.

79. Action to Control Cardiovascular Risk in Diabetes (ACCORD). ClinicalTrials.gov. Available at: http://www.clinicaltrials.gov/ct/show/NCT00000620. Accessed October 5, 2004.

80. Cannon CP, Braunwald E, McCabe CH, et al, for the Pravastatin or Atorvastatin Evaluation and Infection Therapy-Thrombolysis in Myocardial Infarction 22 investigators: Intensive versus moderate lipid lowering with statins after acute coronary syndromes. *N Engl J Med* 2004;350:1495-1504.

81. Smith SC, Blair SN, Borow RO, et al: AHA/ACC guidelines for preventing heart attack and death in patients with atherosclerotic cardiovascular disease: 2001 update. A statement for healthcare professionals from the American Heart Association and the American College of Cardiology. *Circulation* 2001;104:1577-1579.

Appendix A: Case Reports

In July 2004, the National Cholesterol Education Program (NCEP) issued a report updating the 2001 Adult Treatment Panel III (ATP III) guidelines, based on the results of recent clinical trials. The report places an increased emphasis on treating high risk, rather than just high cholesterol, by advocating a low-density lipoprotein cholesterol (LDL-C) goal of <70 mg/dL in patients deemed to be at very high risk, even if their LDL-C levels are already low. Other modifications include earlier initiation of therapy to decrease elevated triglyceride (TG) levels and to raise low levels of high-density lipoprotein cholesterol (HDL-C) in selected patients. The following case reports illustrate how the modified ATP III guidelines may be applied in clinical practice.

In the prevention and treatment of coronary heart disease (CHD), patient compliance is essential. Therefore, it is crucial to involve patients actively in their own care by informing them of their lipoprotein levels and by using multiple compliance strategies and tools, some of which are discussed in Appendices B and C.

Case 1: Elevated Low-Density Lipoprotein Cholesterol in a Patient Who Underwent Coronary Artery Bypass Graft Surgery

A 57-year-old man came for routine follow-up after coronary artery bypass graft (CABG) surgery. He had a history of hypercholesterolemia and hypertension, and his father had suffered a myocardial infarction (MI) at age 62. The patient was a nonsmoker. His medications were aspirin 325 mg/d, atenolol (Tenormin®) 100 mg/d, amlodipine (Norvasc®) 10 mg/d, and atorvastatin (Lipitor®) 10 mg/d. Six months earlier, after presenting with unstable angina, he underwent coronary angiography and was found to have three-vessel coronary artery disease. Coronary artery bypass graft surgery was performed, with a graft of the left internal mammary artery to the left anterior descending artery and two saphenous vein grafts to the distal right coronary artery and the first obtuse marginal branch of the circumflex coronary artery.

The patient, a nonsmoker, had previously been treated at another institution, and this was his first post-CABG follow-up visit. His fasting lipoprotein profile indicated a total cholesterol (TC) of 255 mg/dL, an LDL-C of 180 mg/dL, a TG of 113 mg/dL, and an HDL-C of 33 mg/dL. The patient was 5 ft 11 in tall and weighed 170 lb, his blood pressure was 135/76 mm Hg, and his fasting plasma glucose concentration was 83 mg/dL.

Questions

1. *What is this patient's CHD risk category?**
2. *What would you recommend for this patient?*
   a. Addition of sustained-release nicotinic acid (niacin) (Niaspan® Extended Release), with titration up to 2,000 mg/d, as needed, to raise his HDL-C level
   b. Titration of atorvastatin to 40 mg/d, plus gemfibrozil (Lopid®) 600 mg b.i.d. to raise HDL-C
   c. Titration of atorvastatin to 80 mg/d, with the addition of a bile acid-binding resin or ezetimibe (Zetia®) to enhance LDL-C lowering, if needed
   d. Consideration of LDL apheresis if the LDL-C goal cannot be achieved with drug therapy.

Answers

1. *Risk category 1 (CHD or CHD risk equivalent)*
2. *c*

Case Analysis

Based on this patient's age, recent CABG surgery, markedly elevated LDL-C, and low HDL-C, the physician considers his 10-year risk for a CHD event to be very high. Therefore, according to the revised ATP III guidelines, an LDL-C goal <70 mg/dL is a therapeutic option.

At the first visit, the patient was strongly advised about therapeutic lifestyle changes (TLC); referred to a cardiac rehabilitation program; and seen by a dietitian, who recommended strategies that included the use of plant stanol/sterol-fortified foods. In addition, the atorvastatin dose was increased to 40 mg/dL. By the 6-week follow-up visit, the patient's LDL-C level had decreased to 140 mg/dL. The physician titrated atorvastatin to the maximum approved dose of 80 mg/dL. Six weeks later, a fasting lipoprotein profile showed an LDL-C level of 115 mg/dL. After determining that the patient had been compliant with his statin regimen, the physician decided to intensify LDL lowering through the addition of either a resin or ezetimibe.

Before initiating combination therapy, the following issues should be considered: patient compliance with current monotherapy; therapeutic goal; mechanisms of each drug (different, preferably complementary); minimization of potential negative effects (eg, increase in TG levels) or adverse effects; dosing and ease of administration, and the likelihood of compliance.

Resins or ezetimibe, a cholesterol absorption inhibitor, can both be used in combination with a statin to improve cholesterol lowering. Based on clinical experience, a resin added to statin therapy can reduce LDL-C by 10% to 20% more than statin therapy alone. Resins can also raise HDL-C levels by 3% to 5%, which would benefit this patient because his HDL-C is <40 mg/dL. Ezetimibe in combination with atorvastatin has been shown to lower LDL-C by 50% to 60%, depending on the statin dose, compared with reductions of 35% to 50% using atorvastatin alone (average incremental reduction of 12% with combination therapy). In addition, ezetimibe appears to raise HDL-C by an additional 1% to 5%. Both classes of drugs work in the small intestine, and their mechanisms of action, although different from one another, are complementary to the mechanism of statins. However, resins tend to produce gastrointestinal (GI) discomfort and can interfere with the absorption of other medications (colesevelam [WelChol®], the newest agent in this class, is better tolerated). Moreover, the use of a resin involves either mixing powders in liquids or taking multiple tablets daily. Ezetimibe has greater ease of administration and is not associated with GI side effects. Therefore, based on the above criteria for selecting an adjunctive agent, the physician prescribed ezetimibe 10 mg/d for use with atorvastatin 80 mg/d. The importance of adherence to diet and exercise recommendations was also stressed, because poor diet and a sedentary lifestyle can undermine the effectiveness of drug treatment.

For high-risk patients with elevated LDL-C and low HDL-C levels, the addition of a fibrate or nicotinic acid to raise HDL-C can be considered. Because this patient has been referred to a cardiac rehabilitation program, however, his physician first chose to see if continued exercise would help raise his HDL-C (in addition to the modest increase expected with ezetimibe). If, after a 6-week trial of combination therapy, the HDL-C does not reach 40 mg/dL, the cautious addition of a fibrate or nicotinic acid can be evaluated.

Based on current guidelines, this patient is not a candidate for LDL apheresis (see Case 5).

Case 2: Ten-Year CHD Risk <10% in a Patient With a Strong Family History of Premature CHD

A 41-year-old man came for a routine physical examination. He had no significant medical history, but described a positive family history of CHD. His father had an MI at age 53 and subsequently underwent CABG surgery. Two paternal uncles also had coronary artery disease. The patient, a nonsmoker, was 5 ft 10 in tall, weighed 190 lb, and met the criterion for abdominal obesity according to ATP III guidelines. He was not taking any medications and had no symptoms of cardiovascular disease. Physical examination was unremarkable, and his blood pressure was 136/68 mm Hg. The patient's fasting lipoprotein profile revealed a TC of 210 mg/dL, an LDL-C of 145 mg/dL, a TG of 165 mg/dL, and an HDL-C of 31 mg/dL. His non-HDL-C level was 179 mg/dL.

Questions

1. *What is this patient's CHD risk category?*
2. *What is this patient's Framingham risk score (10-year estimated risk)?*
3. *What would you recommend for this patient?*
   a. A fibrate to raise his low HDL-C level
   b. Reassurance that his lipoprotein profile is reasonable and that he does not need lipid-lowering therapy at his age
   c. Statin therapy to lower his elevated LDL-C level
   d. TLC (diet and exercise) to raise his HDL-C level, possibly lower his LDL-C level, and reduce his obesity, with drug therapy if the fasting lipoprotein profile does not substantially improve in 3 months.

Answers

1. *Risk category 2 (≥2 risk factors)*
2. *8 (4%): moderate risk*
3. *d*

Case Analysis

According to the guidelines, this patient has two major risk factors that may modify the approach to reaching his ATP III LDL-C goal of <130 mg/dL. His HDL-C level is 31 mg/dL, which is below the risk factor threshold of 40 mg/dL, and he has a positive family history of premature CHD. Although the Framingham scoring system does not include family history as a determinant of risk, recent prospective data from the Framingham Offspring Study show an independent association between validated premature parental cardiovascular disease and a twofold increase in cardiovascular disease risk among middle-aged men who were free of clinical cardiovascular disease at baseline. In addition, other factors (eg, certain key features of the metabolic syndrome; selected novel risk factors such as C-reactive protein) can signify increased CHD risk.

This case highlights the importance of clinical judgment. Because the patient has a family history of premature CHD, as well as metabolic characteristics not included in the risk-estimation formula (eg, abdominal obesity, borderline high TG level), his 10-year risk may actually be 10% to 20% or higher, rather than the Framingham-estimated 4%. Therefore, the physician will initiate lipid-lowering therapy if the LDL-C level remains ≥130 mg/dL after a 12-week trial of lifestyle measures. According to the ATP III guidelines, ≥130 mg/dL is the LDL-C cut point for considering drug therapy in the presence of ≥2 risk factors and a risk estimate of 10% to 20%.

At the first visit, calorie reduction and physical activity were emphasized to help achieve a more appropriate weight of 155 lb. Physician encouragement is essential because it can help motivate adherence to dietary and exercise recommendations, possibly avoiding the need for medication. If the LDL-C goal of <130 mg/dL is not achieved by the 6-week follow-up visit, more stringent nutritional measures will be recommended, including the addition of plant stanols/sterols and referral to a registered dietitian or nutritionist. Another 6-week visit will be scheduled.

In addition to prescribing statin therapy (at a dose to be determined) if the LDL-C level remains ≥130 mg/dL after 12 weeks of TLC, the physician will reevaluate the patient's TG and HDL-C levels. If the TG concentration remains elevated and the HDL-C level has not increased, the patient will be referred for a fasting plasma glucose test to assess the presence of elevated fasting glucose with possible insulin resistance. This could confirm the presence of the metabolic syndrome or diabetes. The cautious addition of a fibrate will also be considered to help lower TG and raise HDL-C levels.

Although antihypertensive medication is not recommended now, routine blood pressure monitoring is necessary to detect any increase that may require treatment in the future.

Case 3: Elevated Levels of LDL-C and TG in a Patient With Premature CHD and Diabetes Mellitus

A 34-year-old man came for follow-up after CABG surgery that was performed when coronary angiography subsequent to a lateral wall MI revealed severe two-vessel coronary artery disease and mild left ventricular (LV) dysfunction. He also had a history of elevated blood cholesterol and poorly controlled type 2 diabetes mellitus.

The patient, a nonsmoker, had a 5-year history of diabetes and a very strong family history of premature CHD. He was 6 ft 1 in tall, weighed 232 lb, and his waist circumference met the standard for abdominal obesity according to the ATP III guidelines. His cardiovascular medications were aspirin 325 mg/d, atenolol 100 mg/d, amlodipine 5 mg/d, and atorvastatin 20 mg/d. For treatment of diabetes, the patient was taking metformin (Glucophage®) 500 mg b.i.d. and glipizide (Glucotrol®) 5 mg/d. However, these medications provided inadequate glycemic control. His blood pressure was 145/88 mm Hg. Physical examination was unremarkable, and laboratory tests showed normal thyroid function. The patient's fasting lipoprotein profile revealed a TC of 390 mg/dL, an LDL-C of 180 mg/dL, a TG of 723 mg/dL, and an HDL-C of 22 mg/dL. His non-HDL-C level was 368 mg/dL. Based on the patient's family history and elevated lipid levels, a diagnosis of familial combined hyperlipidemia was suggested.

Questions

1. *What is this patient's CHD risk category?*
2. *What would you recommend for this patient?*
   a. A fibrate to decrease his TG and raise his HDL-C levels
   b. Titration of atorvastatin up to 80 mg/d, as needed, to decrease his LDL-C level, with the addition of ezetimibe, as needed, for further LDL-C reduction, and a fibrate (eg, fenofibrate [TriCor®]) to lower his TG level and raise his HDL-C concentration

c. Nicotinic acid (niacin) alone to lower his TG level

d. Addition of nicotinic acid to atorvastatin.

Answers

1. *Risk category 1 (CHD or CHD risk equivalent)*
2. *b*

Case Analysis

This patient illustrates the considerable problem of severe combined hyperlipidemia at a young age. He is diabetic, overweight, and hypertensive and has a significant history of premature CHD and recent CABG surgery. In addition, he has a family history of premature CHD. Several approaches can be adopted to manage his hyperlipidemia. Aggressive LDL-C lowering to <70 mg/dL, the optional goal in very high-risk patients, is clearly needed. Therefore, it is appropriate to titrate the atorvastatin to a maximum of 80 mg/d and assess the response after 6 weeks. The patient also has a TG level ≥500 mg/dL, placing him at high risk for acute pancreatitis, and markedly low HDL-C in the context of poorly controlled type 2 diabetes. High TG and low HDL-C levels are the most common pattern of dyslipidemia associated with type 2 diabetes.

It is often advisable to refer young patients with diabetes to an endocrinologist. For many patients with type 2 diabetes, insulin therapy can improve glycemic control and help lower TG levels. Alternatively, the regimen of oral agents may be modified by adding a glitazone drug. This patient was referred to a specialist and achieved optimal control of his diabetes with insulin therapy. However, his TG concentration remained consistently >400 mg/dL, and his HDL-C level remained <40 mg/dL. Therefore, fenofibrate 54 mg/d was added to atorvastatin. Fenofibrate is a fibric acid derivative (fibrate) that can be titrated up to 160 mg/d. Fibrates lower TG and raise HDL-C concentrations by decreasing hepatic synthesis of TG; stimulating lipoprotein lipase, which catalyzes the metabolism of TG; and increasing the production of apolipoprotein A-I, a principal component of HDL. Within 6 weeks, the combination of atorvastatin 80 mg/d and fenofibrate 54 mg/d resulted in an LDL-C level of 115 mg/dL, a TG level of 232 mg/dL, and an HDL-C level of 31 mg/dL. Although atorvastatin 80 mg decreased this patient's LDL-C level by approximately 35%, the physician chose to add ezetimibe 10 mg in an effort to reach the optional goal of <70 mg/dL.

Because high-dose atorvastatin and fenofibrate can interact, this patient should be monitored frequently for abnormal liver function and evidence of myopathy.[**,†] Although combination drug therapy is required to control dyslipidemia in this case, the importance of diet and exercise should not be overlooked. After completing cardiac rehabilitation therapy, the patient was prescribed a vigorous home exercise program, which can help raise his HDL-C level, and was referred to a registered dietitian for medical nutrition therapy.

Nicotinic acid is a reasonable alternative to fibrate therapy for lowering TG and raising HDL-C levels. However, in this patient, the presence of diabetes made a fibrate a better choice.[**] Fish oil capsules can be considered in order to improve TG lowering if the level remains ≥150 mg/dL with fibrate therapy.

To reach a blood pressure goal of <130/<80 mm Hg, as recommended by the American Diabetes Association, the addition of an angiotensin-converting enzyme inhibitor is reasonable.

Case 4: Primary Prevention in a Patient With Diabetes Mellitus

A 38-year-old man came for a routine physical examination and advice on cholesterol management. He had a history of type 1 diabetes from age 15 and had been very tightly controlled over the intervening years on combined short- and long-acting insulin. The patient's home glucose monitoring and HbA_{1C} values indicated very good glycemic control. Occasionally, the patient experienced hypoglycemic episodes, usually associated with missed meals. There was no family history of premature vascular disease. The patient, a computer programmer, was a nonsmoker. His physical examination was normal, with a blood pressure of 136/68 mm Hg. Recent ophthalmologic examination revealed minimal background diabetic retinopathy. Studies for urinary microalbuminuria were negative, and thyroid function studies were also normal. His only medications were insulin, folic acid, and a multivitamin. A fasting lipoprotein profile showed a TC of 197 mg/dL, an LDL-C of 145 mg/dL, a TG of 147 mg/dL, and an HDL-C of 30 mg/dL. The patient had seen a recent commercial on television and wanted to know if he was eligible for cholesterol-lowering therapy.

Questions

1. *What is this patient's CHD risk category?*
2. *What would you recommend for this patient?*
    a. Reassurance that he has no vascular disease and that an LDL-C level <160 mg/dL is acceptable
    b. TLC
    c. A fibrate to raise his HDL-C level
    d. A statin to lower his LDL-C level

Answers

1. *Risk category 1 (CHD or CHD risk equivalent)*
2. *d*

Case Analysis

Although this patient does not have a history of CHD, the ATP III guidelines identify diabetes as a *CHD risk equivalent*. This qualifies the meaning of the phrase *primary prevention*, because patients with CHD risk equivalents require the same intensive treatment as those with established CHD. The patient also has low HDL-C and systolic hypertension (according to American Diabetes Association criteria). Based on these multiple risk factors, the physician deems the optional LDL-C goal of <70 mg/dL to be appropriate.

It is unlikely that TLC alone will lower LDL-C from 145 mg/dL to <70 mg/dL. Therefore, simvastatin (Zocor®) 40 mg/d was prescribed, and the patient was asked to return in 6 weeks to monitor his progress. Simultaneous diet and exercise are also important. Although the patient's glycemic control has been good, his sedentary occupation may place him at risk for weight gain and an associated increase in TG levels. The physician prescribed an exercise program, which can also help raise HDL-C and perhaps lower LDL-C. The patient misses meals, suggesting that his diet may not be optimal and that referral to a registered dietitian should be considered. Physician encouragement is needed to promote adherence to dietary and exercise regimens that can maximize the effect of lipid-lowering therapy and help avoid future cardiovascular complications. Although not prescribed now, antihypertensive medication may be warranted in the future if the patient's blood pressure begins to rise. Therefore, routine monitoring is necessary.

Because this patient has early diabetic microvascular disease (ie, retinopathy), referral to an ophthalmologist specializing in the treatment of patients with diabetes is recommended.

Case 5: Isolated Elevated LDL-C in an Asymptomatic Young Woman

A 22-year-old woman was referred to a lipid specialist for evaluation of high cholesterol. She had no medical problems and was asymptomatic and a nonsmoker. Her height was 5 ft 5 in and her weight was 125 lb. Her father had an MI at age 50, but her mother and siblings had no history of cardiovascular disease. She was single and was not taking oral contraceptives. The patient, a vegetarian, reported exercising five times per week. According to the results of a fasting lipoprotein profile, she had a TC of 323 mg/dL, an LDL-C of 208 mg/dL, TG of 185 mg/dL, and an HDL-C of 84 mg/dL. Her blood pressure was 118/60 mm Hg.

Questions

1. *What is this patient's CHD risk category?*
2. *What would you recommend for this patient?*
   a. Vigorous TLC in an attempt to achieve an LDL-C level <190 mg/dL
   b. Lifelong statin therapy
   c. LDL apheresis to lower LDL-C level
   d. Ultrafast computed tomography (CT) scan to assess for coronary calcification for more precise risk stratification.

Answers

1. *Risk category 3 (0 to 1 risk factor)*
2. *a*

Case Analysis

According to ATP III guidelines, the patient's positive family history of premature cardiovascular disease is a major CHD risk factor. As stated earlier, however, family history is not one of the risk factors used by the Framingham scoring system to estimate CHD risk. Furthermore, ATP III guidelines stipulate that an HDL-C level ≥60 mg/dL is a negative risk factor, leaving this patient with a net of 0 risk factors, which suggests a 10-year risk of <10%. Nevertheless, the patient's high level of LDL-C (≥190 mg/dL) calls upon the physician to use clinical judgment in determining whether lipid-lowering drug therapy is needed. Although statin therapy might seem like a reasonable choice at first, it is contraindicated during pregnancy and lactation. Consequently, vigorous lifestyle changes are preferable as first-line therapy in a woman of childbearing age.

Because she is a vegetarian, the patient's daily consumption of total fat and saturated fat is already low. However, it is important to discuss nutrition in detail to determine if she adheres to ATP III dietary recommendations regarding intake of carbohydrates, fiber, and protein. For example, carbohydrates, which can increase TG concentrations, should not exceed 50% to 60% of total calories and should be consumed in the form of whole grains and fresh fruits and vegetables (complex carbohydrates), rather than baked goods or commercially processed foods.

According to ATP III guidelines, this patient's TG level (185 mg/dL) is in the borderline high range (150 to 199 mg/dL). However, available clinical trial data do not support the use of fibrates to lower TG levels in patients with her baseline lipid profile. The physician therefore decided to focus on lifestyle measures, including carbohydrate intake and the possible inclusion of flaxseed oil to provide the ω-3 fatty acids that are supplied by fish consumption in the nonvegetarian diet. To assess the need for these and other nutritional measures and to monitor compliance, the physician may refer the patient to a registered dietitian or nutritionist for medical nutrition therapy.

If a 12-week trial of lifestyle changes does not result in an LDL-C level <160 mg/dL, statin therapy will be prescribed. Because of the risk posed by statins during pregnancy and lactation, however, the patient should be advised to practice contraception while taking a drug of this class. In the future, if she wishes to have a child, she should stop statin therapy before becoming pregnant.

This highlights the importance of effective physician-patient communication. The rapport between physician and patient can have a direct effect on compliance with treatment. In addition, it is important for the physician to learn about beliefs and lifestyle habits that can affect the patient's ability or willingness to adhere to therapeutic guidelines.

In a young patient with markedly elevated cholesterol levels and a history of premature parental CHD, a familial pattern of elevated risk is strongly suggested. Accordingly, the physician should also recommend cholesterol testing of the patient's first-degree relatives.

This patient is not a candidate for LDL apheresis, according to procedures approved by the United States Food and Drug Administration (FDA). Following a 6-month trial of dietary therapy and maximum tolerated combination therapy, LDL apheresis may be appropriate for patients with homozygous familial hypercholesterolemia (FH) and LDL-C levels ≥500 mg/dL; patients with heterozygous FH, failure of medical therapy, and LDL-C levels ≥300 mg/dL; or patients with heterozygous FH, established coronary artery disease, failure of medical therapy, and LDL-C levels ≥200 mg/dL.

A CT scan for coronary calcification is not recommended for this patient because the outcome, whether positive or negative, would not clearly resolve the dilemma of whether to recommend a statin as first-line therapy in a young woman of childbearing age.

Case 6: Emerging Risk Factors in a Patient With a Family History of Premature CHD

A 39-year-old man came for cardiovascular evaluation because of a strong family history of premature coronary artery disease. His father and his paternal grandfather each died suddenly of an MI at age 40 and age 48, respectively. His father's brother had developed premature coronary artery disease, undergone CABG surgery at age 50, and died of complications of coronary artery disease at age 58.

The patient, a nonsmoker, was asymptomatic and worked out regularly in a gym. His height was 6 ft 2 in, his weight was 183 lb, and his blood pressure was 128/80 mm Hg. Physical examination was unremarkable, and the results of a treadmill exercise text were negative. Based on a fasting lipoprotein profile, the patient's lipid values were TC 176 mg/dL, LDL-C 109 mg/dL, TG 132 mg/dL, and HDL-C 41 mg/dL. His fasting plasma glucose was 94 mg/dL. Although the patient's lipid values were relatively normal, his family history indicated elevated cardiovascular risk, which prompted a search for the presence of emerging risk factors that could help determine his individual risk. Tests revealed that his lipoprotein(a) (Lp[a]) level was elevated at 136.9 mg/dL (normal range 1 to 20 mg/dL). Homocysteine was normal at 8.9 µmol/L, and high-sensitivity C-reactive protein (hsCRP) was 1.5 mg/L (indicative of moderate risk). After a detailed discussion of the elevated Lp(a), the intermediate-level hsCRP, and the lack of efficacious drugs to lower Lp(a) in the context of relatively normal lipid values, the patient was sent for an ultrafast CT scan of the heart as a 'tie-breaker test.' This test indicated a calcium score of zero.

Questions

1. *What is this patient's CHD risk category?*
2. *What would you recommend for this patient?*
   a. Primary prevention with a statin to lower his LDL-C level to <100 mg/dL
   b. TLC
   c. Low-dose aspirin therapy because elevated Lp(a) may induce a prothrombotic state
   d. Primary prevention with niacin to raise his HDL-C level.

Answers

1. *Risk category 3 (0-1 risk factor)*

Based on his LDL-C level and his Framingham risk estimate, this patient's likelihood of developing CHD appears low. However, his actual risk may be higher because premature parental cardiovascular disease, his principal risk factor, is not included in the Framingham calculation. This limitation of the Framingham system highlights the need for global risk assessment in all patients. Clinical judgment is crucial in determining when individual risk exceeds the Framingham estimate because treatment must be based on high risk, not just high LDL-C level.

2. *c*

Case Analysis

This case, which highlights the use of emerging risk factors and new technologies to refine risk assessment, illustrates the dilemmas that these investigations can present. Although the patient was asymptomatic, with satisfactory lipid and glucose levels and a negative stress test, the elevations in Lp(a) and hsCRP levels increased his anxiety about his cardiovascular risk. For the physician, this represents an increasingly frequent issue in medical practice: what is the appropriate action when diagnostic tests indicate an abnormality that may signify a disease process, but there are no clinical symptoms and the abnormality is not readily treatable?

In this instance, the ultrafast CT of the heart was useful because a zero calcium score makes the presence of significant subclinical coronary atherosclerosis unlikely. Based on the results of the ultrafast CT, aggressive lipid lowering with an LDL-C goal of <100 mg/dL was not recommended. However, the patient was prescribed aspirin 81 mg/d. Because Lp(a) appears to have an atherothrombogenic effect, it has been suggested that the antiplatelet mechanism of aspirin may be beneficial in a patient with elevated Lp(a) and normal LDL-C levels. Both aspirin and statins have been shown to reduce CHD risk in patients with increased levels of hsCRP, and a large trial is now under way to evaluate whether reducing CRP levels with statin therapy will also decrease cardiovascular risk (see Chapter 5 for a discussion of the JUPITER trial).

If the ultrafast CT had produced a significant calcium score, even in the presence of a normal stress test, the physician would adopt a more aggressive secondary-prevention approach, acknowledging the calcium score as a surrogate for established coronary atherosclerosis. In that case, statin therapy would be recommended.

This patient's daily routine included an adequate level of physical activity. As part of primary prevention, however, the physician also discussed diet, emphasizing the importance of maintaining an appropriate body weight and adhering to the nutritional recommendations of the NCEP and the American Heart Association.

Because of this patient's family history, monitoring at regular intervals is important. Therefore, a 6-month appointment was scheduled to repeat the fasting lipoprotein, Lp(a), and plasma glucose tests. At that time, if levels have not increased (or decreased, in the case of HDL-C), the patient will be asked to return in 6 months for lipid monitoring and a CT scan, followed by an ongoing schedule of 6-month appointments. If lipid levels begin to change in an unfavorable direction, or if subclinical atherosclerosis is detected, drug therapy to lower LDL-C and/or TG levels or to raise HDL-C will be considered, along with more frequent monitoring. At each visit, adherence to dietary recommendations will be discussed.

The physician also inquired about the patient's siblings and cousins with the same familial risk factor, urging that they seek preventive medical care promptly.

Case 7: Statin Intolerance in a Patient With Elevated LDL-C Level and Prior Myocardial Infarction

A 67-year-old man was referred for management of dyslipidemia after a recent MI. The patient, a nonsmoker, had a 20-year history of dyslipidemia and a 7-year history of hypertension and gout. In addition, he had two siblings with coronary artery disease and diabetes mellitus. Three months before evaluation, the patient had experienced an uncomplicated inferior wall MI that was treated with primary angioplasty at another hospital. At the time of evaluation, he was asymptomatic and had gradually resumed physical exercise. His daily medications were aspirin 325 mg, quinapril 20 mg, metoprolol XL 100 mg, allopurinol 100 mg, folic acid 1 mg, and tamsulosin 0.4 mg for benign prostatic hyperplasia. The patient was 6 feet tall and weighed 195 lb; his blood pressure was 155/90 mm Hg. Physical examination was unremarkable. A fasting lipoprotein profile revealed a TC concentration of 245 mg/dL, an LDL-C of 177 mg/dL, a TG of 132 mg/dL, and an HDL-C of 41 mg/dL. Fasting plasma glucose was 82 mg/dL, and both liver function and creatine kinase (CK) tests were normal. Additional history revealed that the patient had been treated with niacin approximately 15 years earlier but could not continue because of side effects.

In an attempt to control his LDL-C level, the patient's family physician had tried a number of statins over the previous 5 years, but each had been discontinued because of muscle symptoms or an inadequate response. Although he had not developed CK elevations on any of these drugs, the patient expressed reluctance to take atorvastatin, simvastatin, or pravastatin (Pravachol®) because they had caused muscle cramps at either the 20-mg or 40-mg dose. Fluvastatin had not produced a 'significant impact' on his LDL-C, and lovastatin was not tried.

Question

1. *What is this patient's risk category?*
2. *What would you recommend for this patient?*
   a. Simvastatin 20 mg/d and cholestyramine (LoCholest®, Questran®, Prevalite®) 4 g b.i.d.
   b. Simvastatin 10 mg/d and extended-release niacin 500 mg/d
   c. Simvastatin 10 mg/d and gemfibrozil 600 mg b.i.d.
   d. Ezetimibe/simvastatin (Vytorin™) 10/10 mg/d.

Answers

1. *Risk category 1 (CHD or CHD risk equivalent)*
2. *d*

Case Analysis

This case illustrates a clinical situation in which patients are significantly dyslipidemic but unable to tolerate appropriate doses of statin therapy. The updated ATP III guidelines highlight the option of reducing the LDL-C level to <70 mg/dL in very high-risk patients. It is unlikely that this patient with multiple risk factors could achieve the ATP III goal with a low-dose statin alone.

Traditionally, intensification of LDL-C–lowering therapy involves titrating a statin to the maximum tolerated dose and then adding a bile acid-binding resin (cholestyramine, colestipol [Colestid®], or colesevelam) if the LDL-C level remains above goal. Clinical experience indicates that a resin may produce a 10% to 20% further reduction in LDL-C when administered with a statin. A more recent approach involves the addition of ezetimibe to statin therapy.

Ezetimibe is the first in a novel class of cholesterol absorption inhibitors that impairs dietary and biliary cholesterol absorption at the brush border of the small intestine. Unlike the bile acid-binding resins, ezetimibe does not affect the absorption of fat-soluble vitamins and is not associated with an increase in TG levels. Approved both as monotherapy and in combination with a statin for patients with elevated LDL-C levels, ezetimibe at the recommended dose of 10 mg may provide an alternative for patients who are intolerant of statins or may be prescribed as adjunctive therapy for patients able to tolerate only low statin doses that cannot reduce LDL-C levels sufficiently. Recently, a combination agent (ezetimibe/simvastatin) was approved.

Clinical trial results indicate that ezetimibe is well tolerated, with an incidence of liver enzyme elevations or myopathy similar to that of placebo in patients receiving monotherapy. When ezetimibe is administered with a statin, the incidence of serum transaminase elevations is slightly higher than with statin monotherapy. Therefore, liver function tests are recommended before initiating treatment with ezetimibe plus a statin.

This patient was treated with ezetimibe/simvastatin 10/10 mg. After 6 weeks of therapy, his LDL-C level of 131 mg/dL was still above the goal of <70 mg/dL. Further reduction in LDL-C may be achieved through diet modification (ie, referral to a nutritionist for medical nutrition therapy). The physician may also prescribe a resin or nicotinic acid, although the latter was not well tolerated by this patient several years earlier.

The possible addition of a resin raises questions about efficacy and the potential risks of polypharmacy. It is not known whether a resin added to ezetimibe will increase lipid lowering. Moreover, because a resin can inhibit the absorption of concomitant medications, it should be taken several hours apart from all other drugs. In view of these considerations, the physician could decide that it is not advisable to prescribe a resin. Consequently, an LDL-C level <70 mg/dL may not be realistically achievable in this patient without the use of adjunctive LDL apheresis (see Case 5 for FDA criteria).

As with all patients beginning statin therapy, baseline CK levels were tested and the patient was told to report any muscle ache or weakness. Because he had previously experienced statin-associated muscle symptoms, periodic CK testing (initially at 6 to 8 weeks, then every 3 months thereafter) is necessary. A moderate increase in CK level (ie, 3x to 10x upper limit of normal [ULN]), with or without muscle symptoms, will first require the physician to rule out a possible nonstatin cause (eg, strenuous exercise). If such a cause is identified, the patient will be monitored weekly until symptoms abate. In the absence of a possible nondrug cause, or if symptoms worsen, three options are available: discontinue the statin and then rechallenge, reduce the dose, or change to another statin. If the CK level is >10x ULN, statin therapy will be discontinued immediately until all symptoms abate, at which time the other options will be considered. If the patient reports muscle symptoms with no other apparent cause and no CK elevation, the statin will be discontinued and, once medical concern is allayed, the patient will be rechallenged or switched to another statin, with weekly CK testing and continued monitoring of muscle symptoms.

Notes

* Executive Summary of the Third Report of the National Cholesterol Education Program (NCEP) Expert Panel on Detection, Evaluation, and Treatment of High Blood Cholesterol in Adults (Adult Treatment Panel III). *JAMA* 2001;285:2486-2497.

Grundy SM, Cleeman JI, Merz CNB, et al: Implications of recent clinical trials for the National Cholesterol Education Program Adult Treatment Panel III guidelines. *Circulation* 2004;110:227-239.

** According to the American Diabetes Association, nicotinic acid, which lowers TG and raises HDL-C levels, can worsen glycemic control. Therefore, physicians should exercise extreme caution when deciding how to treat elevated TG or low HDL-C in patients with metabolic disorders characterized by hyperglycemia and/or insulin resistance. In addition, the combination of a statin and a fibrate or nicotinic acid can increase the risk for muscle toxicity associated with statin therapy. Patients taking either drug combination should be periodically monitored for liver function and cautioned to report any signs or symptoms (eg, muscle pain, urine discoloration) immediately. See Chapter 5 for additional information.

† To avoid adverse effects or drug-drug interactions, physicians must carefully read the product information for each agent; take a complete patient history that includes information about current medications, allergies, and previous adverse reactions to drug treatment; and determine baseline hepatic, renal, and thyroid function.

Appendix B: Strategies to Promote Adherence

This appendix lists the strategies recommended by the National Cholesterol Education Program (NCEP) Adult Treatment Panel III (ATP III) to help physicians:

- enhance their own compliance with patient care recommendations;
- foster patient adherence to prevention and treatment plans.

Physicians are encouraged to think creatively to develop easy-to-use and patient-friendly systems that promote adherence to lifestyle and treatment strategies.

Interventions to Improve Adherence

Focus on the Patient

- Simplify medication regimens.
- Provide explicit patient instruction and use effective counseling techniques to teach the patient how to follow the prescribed treatment.
- Encourage the use of prompts to help patients remember treatment regimens.
- Use systems that reinforce adherence and maintain contact with the patient.
- Reinforce and reward adherence.
- Encourage the support of family and friends.
- Increase office visits for patients unable to achieve treatment goals.
- Increase convenience and access to care.
- Involve patients in their care through self-monitoring.*

Focus on the Physician and Medical Office

- Teach physicians to implement lipid treatment guidelines.
- Use reminders that prompt physicians to attend to lipid management.
- Identify a patient advocate in the office to help deliver or prompt care.
- Use patients to prompt preventive care.
- Develop a standardized treatment plan to structure care.
- Use feedback from past performance to foster change in future care.
- Remind patients of appointments and follow up missed appointments.

Focus on the Health-Care Delivery System

- Provide lipid management through a lipid clinic.
- Use
  - case management by nurses;
  - telemedicine;
  - the collaborative care of pharmacists.
- Develop and implement critical care pathways in hospitals.

*See Appendix C: Compliance Pledge

Adapted from Executive Summary of the Third Report of the National Cholesterol Education Program (NCEP) Expert Panel on Detection, Evaluation, and Treatment of High Blood Cholesterol in Adults (Adult Treatment Panel III). *JAMA* 2001;285:2486-2497.

Appendix C: Compliance Pledge

Below is a Compliance Pledge that is part of an American Heart Association tool kit to help promote patient compliance. It is available at: **www. americanheart.org/presenter.jhtml?identifier=101.**

We encourage physicians to use this and other compliance aids and to develop compliance strategies adapted to the patients they serve.

Take Charge of Your Health

My health is important to me and my family. That's why I'm making a commitment to live a healthy lifestyle to do my part to reduce my risk of heart disease and stroke. I'll work closely with my health-care provider to develop and follow the health plan that works best for me. Because I'm responsible for my own health, I pledge to:

- Visit my doctor for regular checkups
- Take steps to quit smoking if I smoke
- Limit my sodium intake
- Limit how much caffeine and alcohol I drink
- Eat more fruits and vegetables and foods high in fiber
- Try new heart-healthy recipes
- Control my weight and blood cholesterol with a low-saturated-fat, low-cholesterol diet
- Be physically active for at least 30-60 minutes, 3-4 times per week
- Take my medication(s) every day exactly as prescribed
- Remember to refill my prescriptions on time
- Ask my family and friends to support me

I also agree to work to reach and maintain the cholesterol, blood pressure, and weight goals agreed upon with my doctor and listed below:

Cholesterol	**Blood Pressure**	**Weight**
Level Now:	Level Now:	Level Now:
Goal Level:	Goal Level:	Goal Level:

Additional recommendations: ______________________________

____

I know that preventing—or treating and controlling—risk factors for heart disease and stroke is important. I'm signing this pledge to vow to myself, my family and friends, and my doctor that I'll do everything I can to become and stay healthy. I want to be healthy to enjoy my life and loved ones for many years.

____

Patient's Signature Date

As your health-care provider, I pledge to work with you to help you reach your health goals.

____

Provider's Signature Date

Index

A

α-linolenic acid 131
α-tocopherol 130
abdominal aortic aneurysm 82, 90, 94
abdominal obesity 77, 78, 81, 85, 100, 101, 127, 173, 258, 260
abdominal pain 189, 201, 205
abetalipoproteinemia 51
acetylcholine 66
Action to Control Cardiovascular Risk in Diabetes (ACCORD) 127
acute coronary syndrome 24, 25, 27, 65, 92, 95, 171, 245
adherence 274
adipogenesis 206
adiposity 122
Adult Treatment Panel I (ATP I) 147
Adult Treatment Panel III (ATP III) 9, 32, 34, 35, 77, 78, 82, 85, 91-93, 100, 102, 106, 116, 120-123, 145, 149, 164, 165, 170, 176, 215, 233, 244, 254, 258, 260, 263, 265, 271, 274
Adult Treatment Panel III (ATP III) (*continued*)
guidelines 20, 32, 75, 107, 128, 145, 149, 166, 271
Advicor® 198, 222
Aggrastat-to-Zocor (A-to-Z) 25, 246
Air Force/Texas Coronary Atherosclerosis Prevention Study (AFCAPS/TexCAPS) 14-16, 32, 50, 104
alcohol 88, 103, 137, 141, 142, 148, 152, 276
allopurinol 270
American Academy of Pediatrics (AAP) 236
American College of Obstetricians and Gynecologists 220
American College of Cardiology 107, 220, 246
American College of Cardiology Evaluation of Preventive Therapeutics (ACCEPT) 111
American College of Sports Medicine 231

American Diabetes Association (ADA) 220, 241, 242, 244, 273
American Geriatrics Society 220
American Heart Association (AHA) 107, 112, 116, 117, 120, 122, 124-130, 152, 215, 220-222, 234-236, 246, 269
  AHA diet 125
  AHA Task Force on Risk Reduction 149
amlodipine (Norvasc®) 255, 260
amlodipine/atorvastatin (Caduet®) 209
anabolic steroids 165, 177
angina 12-15, 24, 62, 66, 105, 255
angiography 29, 255, 260
angioplasty 270
Anglo-Scandinavian Cardiac Outcomes Trial Lipid Lowering Arm (ASCOT-LLA) 14, 22, 95, 172
anticoagulants 133
Antihypertensive and Lipid-Lowering Treatment to Prevent Heart Attack Trial (ALLHAT-LLT) 22, 23
antioxidants 20, 129
anxiety 149
apheresis 257, 264, 266
Apolipoproteins 43, 46, 47, 49-52, 149
  apo A 46, 56
  apo A-I 47, 50-52, 60, 206, 261
  apo A-II 47, 50, 52, 60, 61
  apo A-IV 47
  apo B 50-53, 57, 132, 181, 190, 193, 207
  apo B-100 46, 47, 51, 52, 54, 57, 203
  apo B-48 47, 51, 52
  apo C 51, 52-54
  apo C-I 47, 54
  apo C-II 47, 51-54
  apo C-III 47, 51, 54
  apo E 47, 51-54, 132, 181, 190, 193, 207
arachidonic acid 132
arrhythmias 131
arterial wall thickening 219
arthritis 205
artificial menopause 218
aspirin 133, 204, 205, 255, 260, 267, 268, 270
aspirin/pravastatin (Pravigard™) 209
atenolol (Tenormin®) 255, 260
atherogenesis 8, 43, 56, 62, 63, 192
atherogenic diet 34, 82, 89, 101
atherosclerosis 6, 7, 9, 10, 25, 29-31, 35, 56, 57, 60, 62-64, 66, 79, 86,

atherosclerosis (*continued*)
87, 131, 141, 145, 192, 206, 218, 219, 269
atherosclerotic disease 5, 75, 82, 90-92, 94, 101, 121, 219, 234
atherosclerotic events 130
atorvastatin (Lipitor®) 21-24, 28, 30, 31, 67, 181, 183, 185, 190, 191, 193, 197, 198, 236, 239, 246, 255-257, 260-262, 270
ATP-binding-cassette A1 (ABCA1) gene 61

B

B vitamins 129
β-blockers 81, 88, 104, 175, 177, 190
β-carotene 130, 148
β-sitosterol 129
Benecol® 130
bile acids 45
bile-acid sequestrants 11, 170, 177-180, 182, 184, 188, 190, 233
biliary obstruction 180
bloating 182, 189
blood pressure 80, 85, 92, 100, 101, 111, 116, 120, 122, 219, 234, 235, 260, 262, 277
body mass index (BMI) 33, 35, 143-145, 148, 149, 219, 234, 235
body weight 116, 120, 126
Bogalusa Heart Study 9, 51
breast cancer 225, 226
brown urine 195

C

C-reactive protein (CRP) 24, 25, 145, 192, 267
Caduet® 209
calcium 126
cancer 225
carbohydrate intake 124, 127, 137, 148
cardiac death 11, 131
cardiovascular death 133
cardiovascular disease 6, 76, 92, 93, 95, 103, 110-112, 152, 165, 219, 220, 225, 226, 230, 234, 235, 238, 241, 258, 265
cardiovascular risk factors 143
Cardiovascular Risk in Young Finns study 10
Centers for Disease Control and Prevention (CDC) 220, 231
cerivastatin 27, 195, 208
children 122, 124
cholecystectomy 229
cholesterol absorption inhibitors 84, 170, 178, 198-200, 229, 243, 256
Cholesterol and Recurrent Events (CARE) 13, 16, 17, 27, 28

Cholesterol Lowering Intervention Program (CLIP) 147
Cholesterol Treatment Trialists' Collaboration 29
cholesterol-efflux regulating protein (CERP) 61
cholesteryl ester 44-46, 48, 53, 54, 57, 61
cholesteryl ester transfer protein (CETP) 52-55, 58, 59, 61
cholestyramine (LoCholest®, Questran®, Prevalite®) 11, 177, 180, 182, 184, 185, 188, 270, 271
chronic heart failure 145
chronic liver disease 207
chylomicron remnants 45, 46, 53
chylomicronemia 101, 102
chylomicrons 44-46, 48, 50-54, 56, 189, 202, 206
cirrhosis 186
clofibrate 12, 207, 230
colesevelam (WelChol®) 177, 180, 182, 184, 189, 257, 271
Colestid® 177, 271
colestipol (Colestid®) 177, 180, 182, 184, 185, 188, 271
Collaborative Atorvastatin Diabetes Study (CARDS) 14, 239, 240
combination therapy 197, 208
computed tomography (CT) scan 264, 266-269
congestive heart failure 120, 128
constipation 182, 189, 228, 233
coronary angiography 29, 30, 255, 260
coronary artery bypass graft (CABG) 254, 255, 257, 260, 261, 267
coronary artery disease (CAD) 111, 131, 134, 255, 266, 269
Coronary Artery Disease Reversal (CADRE) 149
coronary calcification 264
coronary death 141
Coronary Drug Project (CDP) 12
coronary event 13, 15, 32
coronary heart disease (CHD) 7-9, 11, 13, 17-23, 29, 31, 32, 34-36, 48, 50, 51, 56, 57, 66, 75, 79, 82, 85, 86, 90-92, 94, 100, 101, 104, 106-109, 112, 116, 121-123, 126, 127, 131, 140-142, 145, 149, 152, 166, 168, 169, 173, 215, 216, 218, 220, 224, 225, 227, 231, 234-238, 254, 255, 257, 260, 261, 263, 267, 268

coronary heart disease (CHD) (*continued*)
  death from 10, 11, 13, 14, 16, 152
  emerging risk factors 34, 141, 145
  global risk assessment 9
  morbidity 36
  mortality rates 9, 10, 14, 16, 36, 141
  premature CHD 77, 79, 82, 86, 90, 92, 94, 102, 106, 257, 258, 260, 261
  risk 7, 8, 11, 13, 34, 126, 257, 265
  risk equivalents 8, 32, 35, 91, 94, 108, 121, 168, 169, 255, 261, 263
  risk factors 9, 76, 77, 90-92, 144, 145, 152, 174, 227, 237
coronary stenosis 29
corticosteroids 81, 88, 104, 165, 175
Coumadin® 190
creatine kinase (CK) 193, 194, 205, 208, 222, 270, 272
Crestor® 190
Cushing's syndrome 88
cyclosporine (Sandimmune®, Neoral®) 183, 187, 194, 201
cytochrome P-2C9 (CYP-2C9) 197
cytochrome P-3A4 (CYP-3A4) 197
cytochrome P-450 (CYP-450) pathway 197
cytokines 63

D

death 11, 133, 142
depression 149
dermatitis 133
diabetes 8, 9, 15, 18, 22, 28, 35, 45, 56, 75, 79-82, 87, 88, 92, 94, 95, 104, 105, 109, 120, 127, 128, 137, 145, 165, 169, 183, 195, 204, 215, 218-220, 231, 235-238, 241, 242, 244, 260, 261, 262, 269
  type 1 236, 244
  type 2 60, 123, 149, 236, 238, 240, 241
diabetic coma 238
diabetic microvascular disease 264
diabetic nephropathy 242
diabetic retinopathy 262
diarrhea 133, 201
Diet and Reinfarction Trial (DART) 133
Dietary Approaches to Stop Hypertension (DASH) 144
dietary counseling 148, 149
Dietary Guidelines for Americans 2000 135

dietary habits 230
dietary recommendations 117, 124
dietary therapy 116, 126, 147, 165, 230, 234, 244
dietitians 116, 138, 139, 148, 171, 231, 236, 244, 262, 265
digoxin (Lanoxin®) 190
diuretics 88, 128, 190
docosahexaenoic acid (DHA) 131, 132
drug therapy 6, 10, 116, 135, 141, 165, 171, 228, 229
dyslipidemia 5, 7, 9, 80, 88, 101, 107, 127, 128, 165, 177, 190, 197, 202, 203, 227, 228, 235, 243, 261, 262, 269

E

eicosapentaenoic acid (EPA) 131, 132
elderly 22, 109, 215, 227, 228, 230, 233
emerging risk factors 82, 89, 90, 93, 100, 101, 108
endothelium-derived relaxing factor (EDRF) 66
erythromycin 194
estrogen 104, 175, 223, 224
Estrogen Replacement and Atherosclerosis (ERA) 223
estrogen replacement therapy (ERT) 223
European Atherosclerosis Society 244
European Task Force 244
exercise 117, 143, 149, 151, 152, 231
ezetimibe (Zetia®) 168, 179, 187, 197-201, 216, 243, 246, 255, 256, 257, 260, 271, 272
ezetimibe/simvastatin (Vytorin™) 200, 201, 208, 270-272

F

factor VII phospholipid complex 207
familial combined hyper-lipidemia 87, 104, 260
familial defective apolipoprotein B 86, 102
familial dysbetalipoprotein-emia 87, 104, 180
familial hypercholesterol-emia (FH) 52, 86, 102, 235, 236, 266
familial hypertriglyceri-demia 87, 104
family history 257, 260, 265
fasting glucose 34
fasting lipoprotein profile 76, 85, 108

fats
- fatty acids 45, 48
- monounsaturated fats 136, 148
- monounsaturated fatty acids 142
- polyunsaturated fats 136, 148
- polyunsaturated fatty acids 142
- saturated fats 124, 126, 127, 135, 136, 139
- saturated fatty acids 142
- *trans*-fatty acids 124, 130, 135, 136
- *trans*-unsaturated fats 124
- *trans*-unsaturated fatty acids 124
- unsaturated fats 135

fatty streak 7, 60, 63, 64
fenofibrate (TriCor®) 186, 201, 205, 206, 208, 260, 261, 262
fiber 125, 136, 236, 244, 265
fibric acid derivatives (fibrates) 11, 33, 84, 171, 174, 175, 177-179, 186, 197, 198, 203, 205-208, 221, 222, 228, 229, 233, 240, 242, 244, 257, 258, 260-263
fibrinogen 69, 207
fish oils 129, 131-133
flushing 183, 204, 229
fluvastatin (Lescol®) 30, 181, 183, 185, 190, 191, 193, 197, 270
foam cells 57, 63, 64
folic acid 129, 145, 263, 270
Food and Drug Administration (FDA) 124, 125, 130, 177, 198, 223, 226, 266
Framingham Heart Study 8, 9, 76, 95, 104, 142, 258, 268
Framingham Offspring Study 258
Framingham Offspring-Spouse Study 92, 148
Framingham risk scoring system 82, 85, 92, 106, 108, 143, 173, 233, 258, 265
Fredrickson classification 101, 102
Fredrickson phenotyping 9, 202
free cholesterol 43
free fatty acids (FFAs) 52, 53, 54, 60, 203, 237
Friedewald equation 76

G

gallbladder disease 186
gallstones 229
gastroesophageal reflux 189

gemfibrozil (Lopid®) 11, 104, 183, 186, 194, 195, 201, 205, 206, 208, 222, 242, 255, 270
GISSI-Prevenzione trial 133
glipizide (Glucotrol®) 260
glitazones 261
Global Registry of Acute Coronary Events (GRACE) 27
glucagon 45
Glucophage® 260
glucose tolerance 229
Glucotrol® 260
gout 183, 205, 229, 269
guar gum 125
guidelines 128

H

Harvard Atherosclerosis Reversibility Project (HARP) 30
Harvard School of Public Health 142
HbA$_{1C}$ 262
Health Professionals' Follow-Up Study 142
Healthy Women Study (HWS) 219
Heart Estrogen/Progestin Replacement Study (HERS) 224
Heart Protection Study (HPS) 8, 18, 20, 21, 28, 32, 165, 193, 222, 223, 227, 231-233, 238, 244, 245
Helsinki Heart Study (HHS) 11, 104
hepatic lipase (HL) 54, 55, 58
hepatitis 179
high blood pressure 173
high-density lipoprotein (HDL) 45, 46, 48, 50, 52-55, 57-61
high-density lipoprotein cholesterol (HDL-C) 8, 11, 13, 15, 16, 18, 33, 35, 53, 56, 57, 60-62, 76, 79-81, 84-88, 91-95, 101, 103-105, 108, 110, 117, 123, 124, 127, 128, 132, 135, 143, 144, 149, 173, 176, 178, 180, 186, 188, 193, 203, 206, 207, 219, 221, 222, 237, 241, 242, 244, 254, 258, 260, 261, 263, 264, 267, 269, 270, 273
high-sensitivity C-reactive protein (hsCRP) 82, 192
HMG-CoA reductase 46, 49, 184, 190
HMG-CoA reductase inhibitors 12, 16, 47, 178, 179, 181, 183, 185, 186, 190-192, 197, 205, 208, 242

Holter monitoring 67
homocysteine 34, 80, 82, 89, 101, 121, 128, 129, 267
Honolulu Heart Program 231
hormone replacement therapy (HRT) 217, 223-226
3-hydroxy-3-methylglutaryl coenzyme A (HMG-CoA) 27, 46, 49, 184
hypercalcemia 133
hypercholesterolemia 8, 11, 30, 68, 86, 130, 145, 147, 192, 198-200, 227, 255
hyperglycemia 79, 127, 183, 243
hyperhomocysteinemia 145
hyperinsulinemia 243
hyperlipidemia 202, 260, 261
hypertension 9, 15, 77, 79, 80, 82, 85, 94, 107, 110, 126-128, 141, 145, 152, 173, 227, 231, 235, 237, 239, 255
hypertriglyceridemia 35, 43, 45, 56, 88, 103, 104, 128, 132, 134, 182, 197, 207, 230, 237, 242, 244
hyperuricemia 183, 205
hypoalbuminemia 128
hypocholesterolemia 51, 80, 151
hypotension 204
hypothyroidism 81, 165

I

impaired fasting glucose 34, 80, 89, 101, 173
impotence 229
Incremental Decrease in Endpoints through Aggressive Lipid Lowering (IDEAL) 168
insulin 45, 204, 219, 237, 261-263
insulin resistance 35, 56, 60, 77, 101, 120, 123, 127, 144, 173, 176, 205, 243
intermediate-density lipoprotein (IDL) 46, 48, 51, 54, 55, 58, 190, 206
International Lipid Information Bureau 244
ischemia 65, 67
ischemic heart disease 62, 133
ischemic stroke 142
isoflavones 129

J

Johns Hopkins Precursors Study 9
JUPITER trial 192, 268

K

kidney dialysis 128
kidney disease 120, 128
Kuopio Atherosclerosis Prevention Study (KAPS) 68

L

L-arginine 145
lactation 265
Lanoxin® 190
lecithin:cholesterol acyltransferase (LCAT) 58, 59, 86
Lescol® 30, 190
lesions 132
  atherosclerotic 9
  sclerotic 7
life-habit risk factors 34, 82, 89, 100, 101, 108, 121, 141, 173
lifestyle therapy 75, 172, 220, 242, 244
lipase 261
lipid hypothesis 10
lipid management clinics 275
Lipid Research Clinics Coronary Primary Prevention Trial (LRC-CPPT) 11
lipid testing 107
Lipid Treatment Assessment Project (L-TAP) 31
lipid treatment guidelines 275
lipid-lowering therapy 15, 28, 29, 31, 36, 47, 67, 107, 127, 147, 165, 171, 172, 197
Lipitor® 21, 67, 190, 236, 255
lipogenesis 45
lipoprotein 30, 43-46, 48, 49, 52-55, 57, 60, 65, 68, 128, 174, 181, 190, 202, 203, 207, 237, 241
Lipoprotein and Coronary Atherosclerosis Study (LCAS) 30
lipoprotein lipase (LPL) 51, 52, 55, 207
lipoprotein(a) [Lp(a)] 34, 46, 48, 56, 57, 80, 82, 89, 101, 121, 128, 193, 207, 267, 269
liver disease 165, 181
LoCholest® 11, 177, 270
Long-Term Intervention with Pravastatin in Ischemic Disease (LIPID) 13-17, 27, 28
Lopid® 11, 104, 194, 222, 255
Los Angeles Veterans Hospital Administration Study 229
lovastatin (Mevacor®) 15, 16, 32, 105, 181, 183, 185, 190-192, 195, 197, 198, 270

low-density lipoprotein (LDL) 43-46, 48, 51, 52, 54-59, 63, 68, 86, 127, 132, 207, 237
low-density lipoprotein cholesterol (LDL-C) 7, 11-16, 18, 20, 21, 23-25, 28, 30-32, 34, 35, 47, 50, 51, 53, 56, 66, 75, 76, 79, 80, 83, 85, 88, 89, 91-93, 100-102, 105, 107, 108, 116, 120, 121, 123, 124, 126, 128, 129, 132, 135, 138-141, 143, 164, 166-168, 170, 173-176, 179-181, 186, 188, 190, 192, 197, 202, 203, 207, 216, 219, 220, 222, 234, 235, 238, 241, 242, 254, 260, 261, 264-266, 268, 270, 271
lupus erythematosus 218
lycopene 130
Lyon Diet Heart Study 142

M

macrolide antibiotics 183, 194
macrophages 55-57, 63, 65, 68
macrovascular complications 127
magnesium 126
medical interview 77, 85, 107
medical nutrition therapy 83, 116, 120, 122, 126-128, 138, 139, 148, 171, 231, 241, 262, 265, 272
menopause 109, 218, 219, 226
Meridia® 144
metabolic syndrome 34, 56, 76, 77, 80-83, 85, 89, 92, 100, 101, 103, 106, 108, 121, 127, 135, 138, 139, 144, 171, 173, 205, 221
metformin (Glucophage®) 260
metoprolol 270
Mevacor® 15, 105, 190
mevalonic acid (mevalonate) 46, 49
microsomal transfer protein (MTP) 51
microvascular disease 237
mitochondria 194
Modification of Diet in Renal Disease (MDRD) 196
Multiple Risk Factor Intervention Trial (MRFIT) 9, 10
muscle pain 195, 273
muscle toxicity 194, 195, 198, 242
myocardial infarction (MI) 11, 12, 14, 16, 17, 19, 20, 22, 23, 25, 30, 31, 62, 66, 67, 90, 104-107, 111, 131, 133, 141, 142, 218, 225,

myocardial infarction (MI) (*continued*)
232, 239, 255, 257, 260, 264, 267, 270
myocardial ischemia 24, 65, 67, 90, 111
Myocardial Ischemia Reduction with Aggressive Cholesterol Lowering (MIRACL) 23-25, 27, 28, 67, 245
myoglobinuria 194
myopathy 27, 171, 179, 183, 193-195, 197, 198, 205, 207, 208, 221, 228, 242, 262, 271

N

National Cholesterol Education Program (NCEP) 5, 8, 31, 32, 75, 78, 111, 116, 117, 122, 128, 143, 150, 170, 215, 244, 254, 269, 273, 274
NCEP guidelines 34, 111, 117, 126
National Heart, Lung, and Blood Institute (NHLBI) 76, 147, 220
nausea 182, 205
nefazodone (Serzone®) 194, 201
Neoral® 194
nephropathy 242
nephrotic syndrome 137
Niacor® 12, 171, 194, 221
Niaspan® Extended Release 171, 194, 221, 255
nicotinamide 202
nicotinic acid (niacin [Niacor®, Niaspan® Extended Release]) 12, 33, 84, 170, 171, 174, 175, 177-179, 181, 183, 185, 194, 198, 201-205, 221, 222, 229, 230, 233, 243, 244, 255, 257, 261, 262, 267, 270, 272, 273
nicotinic acid/lovastatin (Advicor®) 181, 198, 209, 222
nitroglycerin 66
non-high-density lipoprotein cholesterol (non-HDL-C) 174-176
nondrug therapy 116, 135, 137
Norvasc® 255
Nurses' Health Study 142, 224, 226
nutritionists 116, 138, 147, 171, 231

O

ω-3 fatty acids 129, 131, 133, 145
oat products 125
obesity 5, 6, 34, 82, 88, 89, 101, 123, 127, 128, 149, 173, 234, 235, 258, 259
oral contraceptives (OCs) 88, 218, 264

osteoporosis 126
oxidized low-density lipoprotein (OxLDL) 64

P

pancreatitis 102
partial ileal bypass surgery 30
Pathobiological Determinants of Atherosclerosis in Youth (PDAY) 10, 234
patient compliance 138, 147, 148, 150, 177, 188, 203, 265, 266, 274, 275
  Compliance Pledge 275
patient counseling 203
patient education 150
pectin 125
peptic ulcer disease 181, 205
peripheral arterial disease (PAD) 82, 90, 145
peroxisome proliferator-activated receptors (PPARs) 206
phospholipids 43, 45, 46
physical activity 135, 174, 176
physical inactivity 34, 82, 88, 89, 101, 127, 176
physician compliance 148, 150
plant stanols/sterols 129, 135, 138, 139, 168, 256
plaques 7, 64, 67
plasma cholesterol 9
plasminogen 57
plasminogen activator inhibitor-1 (PAI-1) 207
polycystic ovary syndrome 218
polygenic hypercholesterolemia 86, 102
positron emission tomography (PET) 67
Post-Coronary Artery Bypass Graft (Post-CABG) 32
Postmenopausal Estrogen/Progestin Interventions (PEPI) 224
postmenopausal women 218
potassium 126
Pravachol® 13, 66, 190, 231, 270
pravastatin (Pravachol®) 13, 14, 16, 17, 22-24, 27, 28, 30, 31, 66, 68, 181, 183, 185, 190-193, 195, 197, 201, 231, 236, 246, 270
Pravastatin or Atorvastatin Evaluation and Infection Therapy (PROVE-IT) 14, 24, 25, 27, 29, 245
Pravigard™ 209
pregnancy 88, 107, 181, 216, 265, 266
Prevalite® 11, 177, 270
progestational agents 177
progestins 165, 225

Program on the Surgical Control of the Hyperlipidemias (POSCH) 29
proinflammatory factors 89, 101, 145
Prospective Pravastatin Pooling Project 13, 27, 28, 193, 195
Prospective Study of Pravastatin in the Elderly at Risk (PROSPER) 35, 222, 223, 227, 231-233
prostacyclin 59
prostaglandin synthetase 132
protease inhibitor 194
proteinases 63
proteinuria 196
prothrombotic factors 34, 89, 101
pruritus 204
psyllium 125

Q

quality of life 230
Quebec Cardiovascular Study 105
Questran® 11, 177, 270
quinapril 270

R

Reduction of Cholesterol in Ischemia and Function of the Endothelium (RECIFE) 66
renal disease 237, 238
renal dysfunction 128, 137, 165, 181, 186, 208, 222, 228, 242
resins 11, 84, 177, 179, 180, 184, 188-190, 197, 200, 203, 216, 228, 233, 242, 256, 272
retinoids 81, 104, 175, 177
retinopathy 264
Reversal of Atherosclerosis with Aggressive Lipid Lowering (REVERSAL) 30
rhabdomyolysis 27, 186, 194, 195, 208
risk assessment 76, 77, 85, 91, 92, 102, 108-111
risk categories 91
risk classification 108
rosuvastatin (Crestor®) 181, 183, 185, 190, 191, 193, 196, 197, 201

S

salt intake 165
Sandimmune® 194
saphenous vein graft 255
saturated fatty acids 124

Scandinavian Simvastatin Survival Study (4S) 12, 13, 16, 17, 231
Serzone® 194
Seven Countries Study 141
sibutramine (Meridia®) 144
simvastatin (Zocor®) 12, 13, 17, 18, 20, 25, 27, 66, 129, 181, 183, 185, 190-192, 197, 198, 209, 238, 239, 263, 270
sitostanol ester 129
sitosterolemia 131, 200
sleep apnea 128
smoking 5, 9, 28, 77, 82, 88, 92, 94, 95, 107, 137, 141, 145, 151, 152, 169, 173, 176, 218, 219, 227, 231, 234, 235, 238, 239, 276
sodium 128, 276
somatostatin 45
soy proteins 129
stanol/sterol esters 129
statins 11, 23, 24, 27, 28, 36, 47, 49, 66, 69, 84, 107, 168, 170, 171, 178, 179, 185, 187, 190, 192-197, 199-201, 203, 205, 208, 216, 220, 221, 228-231, 233, 239-244, 246, 256, 258, 259, 263-268, 270-272
Step I diet 143, 151
Step I guidelines 143
Step I program 117
Step II diet 143
Step II guidelines 143
Step II program 117
steroids 54, 129, 165
sterols 129
Stockholm Ischaemic Heart Disease Secondary-Prevention Study 12, 229
stool softeners 190
stress 66
stroke 5, 14, 16, 17, 19, 25, 28, 29, 133, 225, 226, 232, 239, 276, 277
Study of the Effectiveness of Additional Reductions in Cholesterol and Homocysteine (SEARCH) 168
subclinical atherosclerotic disease 89
surgical intervention 30

T

Take Control® 130
tamsulosin 270
Tangier disease 61
Tenormin® 255
therapeutic lifestyle changes (TLC) 9, 10, 34, 83, 84, 128, 137-140, 164, 165, 173, 174, 228, 230, 242, 258, 264, 267
  diet 136
thrombogenesis 127, 237
thrombolysis 48, 57
thrombosis 57, 63, 65, 68

thromboxane A_2 132
thyroxin 190
tobacco use 151, 152
total cholesterol (TC) 8, 11-13, 15, 16, 50, 51, 76, 80, 85, 86, 91-93, 105, 108, 116, 128, 129, 132, 143, 144, 188, 198, 231, 237, 241, 255, 258, 260, 263, 264, 267, 270
transaminase elevations 195, 204, 271
transient ischemic attack 17
Treating to New Targets (TNT) 168
TriCor® 201, 260
triglycerides (TG) 8, 11, 12, 15-17, 32, 33, 36, 44-46, 48, 50, 52-54, 56, 59, 60, 76, 79-81, 85, 88, 92, 95, 100, 101, 103, 104, 107, 108, 123, 124, 127, 128, 131, 133, 135-137, 143, 144, 171, 173-175, 178-180, 186, 188, 189, 193, 203, 205, 206, 218, 219, 221, 230, 234, 237, 238, 241, 242, 244, 260, 261, 263-265, 269

U

ubiquinone 194
ultrasonography 29
ultrasound 30
urinary microalbuminuria 263

V

vascular disease 30
vasoconstriction 131, 132
very-low-density lipoprotein (VLDL) 44-46, 48, 51, 53-56, 76, 104, 190, 202, 206, 237
very-low-density lipoprotein cholesterol (VLDL-C) 8, 32, 85, 89, 104, 128, 132, 143, 174, 175, 188, 193, 238
Veterans Affairs High-Density Lipoprotein Trial (VA-HIT) 104
vitamin A 133, 148
vitamin B_{12} 148
vitamin B_6 148
vitamin C 142, 148
vitamin D 133
vitamin E 133, 148
Vytorin™ 200, 270

W

waist circumference 77, 78, 85, 100, 143, 260
warfarin (Coumadin®) 183, 190, 208
weight reduction 174, 176, 244
WelChol® 177, 257

West of Scotland Coronary Prevention Study (WOSCOPS) 13, 16, 27, 28
WISDOM trial 226
women 22, 36, 109, 111, 150, 215-223, 226, 241
Women's Health Initiative (WHI) 225, 226

Z

Zetia® 168, 246, 255
Zocor® 12, 66, 129, 190, 238, 263

NOTES

NOTES

NOTES

NOTES